Instructor's Guide To

Calculus
An Active Approach with Projects

The Ithaca College Calculus Group

Stephen Hilbert
John Maceli
Eric Robinson
Diane Driscoll Schwartz
Stan Seltzer

JOHN WILEY & SONS, INC.
New York • Chichester • Brisbane • Toronto • Singapore

For Alisa, Alison, Elizabeth, Ingrid, J.J., Margaret, Matt, Mike, Monica, Nancy, Peter, Rebecca, Steve, and Sue.

Copyright © 1994 by John Wiley & Sons, Inc.

This material may be reproduced for testing or instructional purposes by people using the text.

ISBN 0-471-03203-4

Printed in the United States of America

10 9 8 7 6 5 4 3 2 1

Introduction

Calculus: An Active Approach with Projects is a collection of materials for first-year calculus developed and tested at Ithaca College. It is not a complete textbook, but a complementary volume that can be used in conjunction with any textbook.

The authors are convinced that students who are actively involoved in class are more likely to succeed than those who are passive. We also view calculus as a unified subject rather than a linearly ordered sequence of topics and believe that this view should be conveyed to students from the outset of their studies. The materials in *Calculus: An Active Approach with Projects* were designed with these beliefs in mind.

There are two main sections in this *Instructor's Guide*. The first section contains activities that can be done in class or as homework. The second section contains large projects for the students to work on (usually in groups) outside the classroom.

Activities

The class activities are designed to accustom students to active participation in the course and to introduce some of the material and methods we have identified as important. Through the activities, students participate in the development of many of the central ideas of calculus. The key is active participation. This kind of student involvement fosters understanding and retention of course material. By doing activities, students also learn modeling and how to use the top-down approach to solve problems—both are useful for successful completion of projects. Activities also help students learn how to draw and interpret graphs—a key element in learning ways to represent functions that are not necessarily given as formulas. Finally, the first set of activities provides an overview of most of first-semester calculus. Doing a number of these activities early in the course helps students see the unity of the subject.

Each activity requires between five and fifty minutes to complete—our estimate for each activity in the book is given in this *Instructor's Guide*. Activities that you schedule to be done in class can, at your option, be done by individual students, or by pairs or groups of students working together. Many are suitable for use as homework assignments. In our courses we like to have an activity in as many classes as possible.

Projects

Projects serve to reinforce material already presented, motivate concepts, or introduce topics that might not otherwise be covered. Most projects involve more than one mathematical concept, and many have open-ended components. We have students work on the projects in teams of three or four, submitting a single report, although many of the projects have parts that are completed by individual students. Most of the projects require about two weeks to complete. In a typical semester course, we have our students do three or four projects.

A significant benefit of this project-oriented approach is that students learn to solve non-trivial, multi-step problems. Working on the (shorter) activities in a guided classroom environment helps them succeed on projects.

"New" Calculus

Efforts to revise the way calculus is taught have focused on a number of different issues. As stated above, the materials presented in *Calculus: An Active Approach with Projects* are designed to empower the student to take an active role in her or his own learning. We emphasize the role of calculus as a tool for understanding the world and hence focus on modeling as a central theme. We also emphasize the notion of function and are careful to show that functions can be represented in many different ways: as graphs, as tables of values, as algebraic expressions, as descriptions (written or verbal), as physical relationships, and as theoretical models.

Course Logistics

Curriculum

Teaching a calculus course using "new" materials usually requires some modification of one's teaching style and some reorganization of topics. As an aid in this adaptation, we have included a *sample* curriculum for a two-semester sequence. The curriculum is offered as an "existence proof" rather than a prescription—there are many possible variations that will result in a successful course based on the materials in *Calculus: An Active Approach with Projects*. The sample curriculum is included as Appendix A in this *Instructor's Guide*.

The sample curriculum illustrates a change in course organization—the spiral approach—that we have found to be particularly effective. Our goal is to present the main ideas of the course early so that the students will see calculus as a unified subject. The emphasis at this stage is on concepts and relationships, not on technical details.

We use the "calculus of graphs" for this purpose. That is, representing the functions involved almost exclusively in graphical form, and using the familiar ideas of velocity and distance as examples, we examine basic ideas from both differential and integral calculus. Within days, the students have some basic understanding about rates and slopes, concavity, and integration (in the context of obtaining a distance graph when given the corresponding velocity graph).

During the rest of the course, students encounter the same ideas again and again, each time picking up more of the technical and computational details.

Exams and Quizzes

Most instructors agree that new approaches to calculus call for new kinds of quiz and examination questions. We have included a sample of the kinds of quiz and examination questions that we have used at Ithaca College in conjunction with our courses that are based on the materials in *Calculus: An Active Approach with Projects*. This collection appears as Appendix B in this *Instructor's Guide*.

Other Issues

Finally, when we have conducted workshops on our active approach to calculus, a number of questions have arisen about how one can integrate these materials into a course. We include some of the most often asked questions and our responses in the next few sections of this introduction.

About Using Projects

Question: How can using projects improve my course?

Answer: Projects bring out both the relevance and the unity of calculus. Most of the projects are set in "realistic" situations. Most also involve more than one calculus topic, often combining topics in unexpected ways. Students need to broaden their perspective and synthesize ideas in order to complete these projects successfully.

Many of the projects have "open-ended" parts—that is, parts for which there is not a unique correct answer or approach. These questions encourage the students to brainstorm in their groups and to view mathematics as a subject with creative elements.

To complete a project, each team needs to submit a well written report of its solution. Writing about mathematics may be a new experience for many students, but it is a valuable one. Describing the solution of a significant problem pricisely in words requires a deeper understanding than most students gain from just solving many problems that are based on examples found in their notes or textbooks.

Question: How do I organize project assignments?

Answer: We believe that the students learn more calculus if they work on the projects in teams. Three students to a group seems to be a good working number, but we have used teams of four or two students. The number depends on the class size and the particular project.

We require each team to meet with the instructor soon after the project is assigned and to present an outline or a top-down analysis of the problem and their anticipated solution at that time. This assures us that the groups have met and that they have begun to think about the problem long before it is due. For a first project, we often require a second meeting to answer questions and get an intermediate progress report.

The final report is expected to conform to guidelines we hand out at the beginning of each semester. A sample is included in Appendix C.

Question: How many projects should I use?

Answer: We have found three or four projects to be an effective choice for a fourteen-week course. We are convinced that assigning several projects provides significantly greater benefit than assigning only one project.

Question: How long do the students work on each project?

Answer: Most projects take two to three weeks for a group of students to complete. The work is done outside class and is in addition to any daily assignments that we make for the class. Our students have a project in progress most of the time.

Question: When should the first project be assigned?

Answer: We believe that it is best to assign the first project early in the course. This helps students to view the projects as an integral part of the course experience. It also helps them to develop significant problem-solving skills that they can bring to bear on other elements of the course.

Question: How should projects affect the course grade?

Answer: Substantially. We count each project as 10% of the total course grade, so that in a four-project class, 40% of the course grade is based on the projects. This is a very effective way to ensure that the students take the projects seriously and devote substantial thought and effort to them.

Question: Can I use projects without having the students work in teams?

Answer: Yes, but at a price. If students are working together to solve these problems, you can have higher expectations for the solutions—after all, they had the advantage of collaborative efforts. More important, working on a team provides effective peer pressure on students not to give up trying to solve a problem. For the instructor, there is the added advantage of fewer projects to grade!

Question: How do I answer questions about a project?

Answer: Just as you would any other graded assignment. Ask them what they have tried and why. Be sure the project team has discussed the question as a group and that everyone on the team has tried to answer it. Then suggest they think about certain ideas that might prove helpful. Try to steer them in appropriate directions without "giving away" solutions. If the same question is raised by a number of teams, it is sometimes helpful to discuss the issue briefly with the whole class.

Question: How much class time should I spend on projects?

Answer: We spend little or none. Our class efforts are devoted to making available to the students the tools that they will need in order to complete the projects successfully. These include the standard tools of calculus, new approaches that can be found in the activities, and practice (again through the activities) in modeling and other problem-solving strategies. One might spend a little class time *after* a project is completed, "debriefing" and discussing how the project leads to or illustrates general concepts.

Question: How do I grade projects?

Answer: That's a good question. In our courses, the team receives a single grade on the group parts of the project. In some cases—for example, "Designing a Detector"—that is the entire project; in other cases—for example, "Spread of a Disease"—there are individual parts that receive a grade separate from the team grade.

It's a good idea to think about how you are going to assign grades before you make the assignment. After you collect the reports, scan all the solutions before you begin grading, and then make a firm decision, again before you begin grading, about how you will determine the grade.

Because there are significant written reports involved, some credit should be allocated to the manner of presentation. Derivations should be explained, graphs and other figures should be labeled, and language should be used correctly.

You need to observe whether they answered the questions posed and how well. Be prepared to reward creative ideas.

The reports should improve throughout the term. When you return first projects, indicate how presentation and reasoning can be improved for future reports. One way to do this is to give out a "collage" of good solution pieces.

Questions About Using Student Groups

Question: How do I form groups? What do I do about students who prefer to work alone?

Answer: For projects, we assign students to groups. The assignments are often based on where people live—for example, students who live in the same dorm will have a good chance of finding convenient ways to meet—or on similarity of schedules. We usually do not ask students who they would like to work with.

We usually change the groups at least once during the semester. This gives the students the opportunity to get to know several others in the class well and gives us the chance to "fix" problem groups.

Sometimes we have project groups work on an activity in class, where we can observe them and make suggestions about procedure and group dynamics.

Occasionally a student expresses the wish to work alone on the projects. Usually, we assign such a student to a group for the first project but permit her or him to work independently on at least one subsequent project.

Question: What do I do about students who don't carry their share of the load?

Answer: There are a number of techniques that we have found effective. First, you should be sure that some of the projects involve individually graded parts. These are usually the data-generating portions of the projects. This assures that a student who is not doing the work will not receive precisely the same grades as those who are.

Second, include a project-related question on the next quiz or exam.

Finally, you can adjust the group membership for subsequent projects. A method that we have found effective is to put the non-workers together in a group. This solution is not intended to be strictly punitive—often such students were simply quiet or intimidated by other group members. Putting such students together in a team often draws them out. This makes the other students feel better and often has the happy result that the members of this new group become motivated to do a good job on the next project.

A related question, of course, is how one identifies students who are not carrying their fair share of the load. One method we have found effective is to require students to submit confidential peer evaluations of all other team members at the end of each project.

Questions About Using Activities

Question: How can using activities improve my course?

Answer: Activities involve the students in their own learning. Both attendance and participation improve. Students who have used activities regularly in class tend to make better comments and ask more significant questions about course material than did those in our more traditional classes.

The activities in *Calculus: An Active Approach with Projects* have several purposes. They introduce new calculus topics, often in a guided discovery format. This reduces the amount of formal presentation that the instructor must do. They help the students become better modelers and problem solvers.

This helps the students complete projects successfully and also helps them solve shorter problems. Students who have experienced activities and projects do not regard "word problems" with dismay.

Question: How do you do an activity in class?

Answer: Each activity is different. Some involve several pieces of student work, each followed by the instructor drawing out solutions and helping with summaries. Others are done more independently by the students. Some are intended to be done outside class as follow-ups to classroom work.

Some of the activities (for example, "Gotcha") are short open-ended problems. Others (for example, "Fundamental Theorem of Calculus") provide a road map to some new calculus concept.

While the students are working, you should watch and listen to some of their discussions. You can give some guidance, but don't be too quick to show them an answer. Sometimes students will help students in another group, and we don't discourage this.

We have found that it is important that the students see the solutions and some sort of summary at the end of each activity. This immediate feedback solidifies their understanding.

Question: Do activities affect the course grade?

Answer: We usually avoid grading activities, preferring that the students feel free to use them to explore calculus ideas without thinking about how the explorations will affect their course grades. However, exams and quizzes do include questions related to the ideas they get from doing the activities.

Question: How long does each activity take?

Answer: A few activities require an entire fifty-minute class period, but many of them are much shorter. Some require as little as five minutes. A discussion of each activity appears in this *Instructor's Guide*. These discussions include estimates of how much time each activity requires.

Questions About Course Organization and Content

Question: How do activities and projects work together?

Answer: Both activities and projects help the students learn modeling and other problem-solving skills, and both force them to be active learners. Working on activities is good preparation for working on projects. Some of the same approaches are appropriate, although the problems in the activities are shorter and done in a more guided setting (the classroom). Working on projects helps the students to become self-starters, which helps them with their work on the activities.

Question: How do you teach problem solving?

Answer: We teach the top-down approach to problem solving. That is, we teach the students to break large problems into successively simpler pieces until the pieces are such that they can find solutions, then to reassemble the pieces into a solution of the original problem. We use this approach in working on problems and activities in class and expect the students to carry the ideas over to the solution of the large problems in their projects.

We also teach them to experiment with special cases, think about similar problems they have seen, make approximations, and use graphs to gain insights about the problem.

Question: How will I have time for activities?

Answer: We have found that incorporating activities into our classes does not result in time pressure. One reason is that many of the activities tend to be fairly short and can be done by the students as they settle in at the start of the class. The other is that the activities actually serve as a means of developing the course material. The student-involved approach replaces some of the formal presentation that we were accustomed to do—and does a more effective job. When the students learn through the activities, they absorb the material better and retain it longer.

Question: How do I present new material?

Answer: Quite a bit of new material comes from the activities, and still more can be introduced in projects. Furthermore, the project/activity approach does not preclude including some "traditional" kinds of presentations. We have found that using activities predisposes our students to participate more fully in even the more traditional classes, resulting in more lively classes and more give and take between students and instructor.

Question: How do quizzes and exams relate to activities and projects?

Answer: We include ideas first encountered in activities and projects on the quizzes and exams. As we mentioned above, this is one way to ensure some degree of participation in project work by all members of the project groups. We also feel free to ask non-standard questions—questions that do not correspond to any template the students have seen—on exams. Such questions are in the spirit of original problem solving that we emphasize through the activities and projects. Some sample questions are in Appendix B.

Question: My students need to have good computational skills. How will they learn to perform computations?

Answer: Much as they always did—through daily homework assignments. There is a difference, however. In the traditional course, students come to believe that calculus *is* computations. In the new course, students see the subject itself— its unity and applications—and regard routine computational problems as an easier (and less interesting) part of the course.

We also still include some rote computations on exams and quizzes, but these are not the main focus. They are necessary tools that the students need to learn.

Question: How do you teach the logical and theoretical aspects of calculus?

Answer: While the theoretical side of calculus is taught less formally than in more traditional models, the emphasis in our courses is on conceptual understanding. We believe this approach lays a more solid foundation for subsequent study than rote memorization of definitions and theorems.

We still teach the formal definition of continuity, the derivative and the definite integral, and significant theorems (the intermediate value theorem, mean value theorem, fundamental theorems, etc.). We emphasize critical thinking about these concepts and understanding their meaning.

Question: Will I be able to cover all of the material?

Answer: We do. The activities and projects serve the same purposes as, for example, several days studying "word problems" and do the job more effectively.

Question: What is the role of technology?

Answer: Several of the activities presuppose the use of either a computer or a graphing calculator. Some of the projects are greatly simplified if some computational device is available to help with the calculations and graphs. The *Instructor's Guide* lists what, if anything, is needed for any particular activity or project. We do not prescribe any particular choice of technology, however. We have always described our materials as *technology independent*. This means that most of the activities and projects require no technology at all, and the few that do require a computer or graphing calculator are presented in such a way that the instructor using the material can choose whatever implementation is available.

Unifying Threads

In our work on calculus, we have identified a number of unifying threads that typically run through a successful first-year course. We have noted with interest that a number of other groups of mathematicians working on revitalizing calculus have identified similar themes. We label our threads **graphical calculus, distance and velocity, multiple representation of functions, modeling, top-down analysis,** and **approximation and estimation**. We give a brief description of each thread below. We describe these themes as "threads" because they are woven throughout the course, and serve to bind it together into a unified whole. Most of the projects, and many of the activities contain elements from more than one of the threads. In Appendix D we have included guides to the relationships between the projects and activities in the volume and the corresponding threads.

Threads in first-year calculus

- **Graphical calculus.** We use graphs as important examples of functions from the start of the first semester of calculus. Many of the important concepts of calculus are presented using graphs of functions (with no formula given) during the first two weeks of the course. These concepts include the slope of a function at a point, increasing and decreasing functions, concavity, continuity, extrema, and the fundamental theorem of calculus, as well as functions paired with their derivatives in the form of graph/rate-graph pairs. We continue to use many examples of functions represented by graphs throughout the year.

- **Distance and velocity.** While graphing (slopes and areas) provides one convenient example of derivatives and integrals, distance and velocity provides another. Velocity is one of the few examples of a rate that students have experienced, so students are able to understand it more intuitively than other examples. It is easy for the instructor to model velocity by simply moving around the classroom. Over the course of two or three days, we are able to introduce the important concepts of calculus by discussing average velocity, instantaneous velocity, and approximations to distance traveled. It is also easy to motivate the first derivative test, for example, by noting that the distance is maximized (locally) "when you turn around"; i.e., when velocity changes from positive to negative. Of course, we don't necessarily refer to derivatives the first day, but when we get there, this result looks familiar.

- **Multiple representations of functions.** We emphasize the notion of a function stressing that there are several representations for a function: graphical, numerical (a table of data), algebraic (a formula or expression), physical, theoretical, and written or verbal. This is important, because students need to be able to apply techniques they learn to functions presented in any of these formats. It is not enough for students to simply learn to manipulate symbols.

    This idea also helps students see the unity of calculus. For instance finding the area under a graph, finding the antiderivative of a function, and computing a Riemann sum from a table of values are all examples of the concept of integration. We emphasize the interplay of all representations. If we have a result for a function some typical questions from the instructor are "What if you were given a graph? Then what does this result tell you? What if you were given a table of values? What does this result mean? Describe this situation in English."

- **Modeling.** Modeling begins early in the course and continues throughout the first year. Projects and classroom examples typically begin with some "real" application, which must be translated into a mathematical model.

  Modeling is introduced early by way of classroom activities. Graphical relationships are the first instance of modeling. For example, in class students are asked to create and analyze a graphical model of the height of a flag from the ground as a function of time as the flag is being raised.

  The students use modeling techniques on the projects. For example, in a Calculus 1 project, students are asked to model the motion for a detector that is guarding a hallway against intruders.

  This emphasis on modeling in class and in the projects seems to have strengthened the students' belief that it is possible to construct functions that model even complex situations and that the concepts presented in calculus are valuable tools in this process. Furthermore, it seems to have significantly reduced the students' fear of word problems.

- **Top-down methodology.** The top-down method is a very effective tool in problem solving. This approach involves breaking a problem into smaller problems and breaking those problems into smaller problems until the resulting problems can all be solved easily. This scheme includes a method for assembling the solutions of the smaller problems into a solution of the original problem.

  We spend some class time introducing the top-down approach to problem solving, and examples in class are approached top down whenever appropriate. ("What things do I need to know? ... Now, what do I need to do in order to know these things? ... ") We emphasize using this approach on any large problem. Most projects are designed to be attacked in a top-down fashion. Students don't need to discover "big" solutions if they can combine small solutions to solve big problems.

- **Approximation and estimation.** Approximation and estimation are stressed as fundamental and unifying concepts in calculus in a theoretical sense (they appear in the concepts of the limit of a function, the derivative, the integral, and the convergence of sequences and series). Moreover, these notions are stressed as appropriate solutions to specific problems. For example, students are asked in class in a group setting to estimate the slope of a graph of a function at a point using only the graph of the function on graph paper and a ruler.

They then compare answers and discuss the question of who has the "correct" answer. Similarly, they experience approximation by applying at least one numerical method to the problem of root finding to a given degree of accuracy and they study at least one method of numerical integration, complete with an error estimate.

Students are also asked to compute Riemann sums and finite differences from tables of values and solve certain problems using these numerical estimates. Indeed, the students view some modeling experiences as estimation or approximation of a "real" situation. They also see Taylor polynomials and/or other functions obtained from curve fitting serving as approximations to the original function, which may be used in place of the original function under certain conditions. Finally, where appropriate, estimation is stressed in connection with common sense. That is, students are asked if their answers to problems make sense. In this way, "mathematical" common sense is both developed and reinforced.

Contents

I Activities 1

1 Graphical Calculus 3
Chalk toss 3
Classroom walk 5
Biking to school 6
Raising a flag 8
Library trip 9
Airplane flight with constant velocity 11
Projected image 15
A formula for a piecewise-linear graph: Top-down analysis 19
Water balloon 22
Graphical estimation of slope 24
Slope with rulers 28
Examining linear velocity 30
Given velocity graph, sketch distance graph 34
Function-derivative pairs 39
More airplane travel 40
Dallas to Houston 43
Water tank problem 45
Tax rates and concavity 47
Testing braking performance 50
The start-up firm 51
Graphical composition 53
The leaky balloon 57
Inverse function from graphs 58

2 Functions, Limits, and Continuity 63
Introduction to functions 63
Postage 66
What's continuity? 68
Limits and continuity from a graph 70
Slopes and difference quotients 72

xv

		Sequences	73
		Can we fool Newton?	75

3 Derivatives — 77

- Linear approximation 77
- Estimating cost 79
- Finite differences 81
- Using the derivative 83
- Gotcha 86
- Animal growth rates 87
- The product fund 88
- Exchange rates and the quotient rule 91
- Using the product rule to get the chain rule 93
- Magnification 95

4 Integration — 99

- Time and speed 99
- Oil flow 101
- Can the car stop in time? 102
- Fundamental theorem of calculus 103
- Comparing integrals and series 105
- Graphical integration 109
- How big can an integral be? 112
- Numerical integration 114
- Verifying the parabolic rule 116
- Finding the average rate of inflation 118
- Cellular phones 120
- The shorter path 121
- The River Sine 124

5 Transcendental Functions — 127

- Ferris wheel 127
- Why mathematicians use e^x 131
- Exponential differences 136
- Inverse functions and derivatives 137
- Fitting exponential curves 140
- Log-log plots 142
- Using scales 144

6 Differential Equations — 149

- Direction fields 149
- Using direction fields 150
- Drawing solution curves 152

CONTENTS   xvii

    The hot potato 153
    Spread of a rumor: discrete logistic growth 155
    Population 157
    Save the perch 158

7 Series   **161**
    Convergence 161
    Investigating series 162
    Space station 163
    Decimal of fortune 165
    Approximating functions with polynomials 167
    Introduction to power series 168
    Graphs of polynomial approximations 169
    Taylor series 172
    Approximating logs 173
    Using series to find indeterminate limits 174
    Using power series to solve a differential equation 175
    Second derivative test 177
    Padé approximation 178
    Using Taylor polynomials to approximate integrals 180
    Complex power series 182

II Projects   **185**
    Designing a roller coaster 187
    Tidal flows 190
    Designing a cruise control 191
    Designing a detector 193
    Taxes 194
    Water evaporation 196
    Mutual funds 197
    Rescuing a satellite 198
    Spread of a disease 199
    Tax assessment 201
    Dome support in a sports stadium 202
    The fish pond 204
    Drug dosage 205
    Investigating series 206
    Topographical maps 207

III	**Appendices**	**209**
A	**Sample Curriculum**	**211**
B	**Sample Questions**	**219**
C	**Guidelines for Projects**	**267**
D	**Guide to the Threads**	**269**

Part I

Activities

Chapter 1

Graphical Calculus

Graphical Calculus and Modeling

The first chapter of this book consists of problems and activities designed to introduce you to many of the important ideas of first-year calculus in a way that encourages you to visualize the objects and actions you are studying. We call this approach the *calculus of graphs*.

We see graphs not only in mathematics, but also in the physical sciences, the social sciences, even the daily newspaper. Graphs are a way of comprehending the world. Graphs give us a way to visualize an ongoing process as a whole. That is, a graph can contain the whole past history of a process, and a prediction of its future progress, in a way that can be comprehended quickly. In short, the usefulness of graphs illustrates the truth of the old saying, "One picture is worth a thousand words."

In many of the activities you will be asked to work with phenomena or functions that have been represented only as graphs, and to sketch graphs to represent real world problems you are studying. In many cases, you will not have a formula that corresponds to the graph—the graph itself contains all the information for the problem. This approach will enable you to see the big picture of calculus, while temporarily postponing many of the technical details.

At the same time, you will get an introduction to *modeling*. Modeling is the process of representing real world problems in mathematical terms so that the methods of mathematics can be used to gain understanding of the original problem. When you sketch a graph based on your observation of some action or from reading a verbal description, you are constructing a mathematical model. The effective use of mathematical models has been responsible for much progress in the physical and social sciences.

The approach to the learning of calculus that we take throughout this book emphasizes both the calculus of graphs to gain understanding and insights, and modeling to help you become a good problem solver.

Chalk toss

Topic: Graphical modeling

Summary: Students observe a physical motion and record the position function as a graph.

Time required: 20 minutes

Threads: graphical calculus, modeling

Calculus is a study of changes. One form of change is the change in the position of something that is moving. For example, if your instructor throws a piece of chalk into the air and then catches it, the distance of the chalk from the floor changes.

We can record the chalk's position relative to the floor on a graph. The graph will capture the information about how the position changes over time.

1. Watch as your instructor tosses the chalk, and record what you see on a graph.

Here you should toss the chalk in the air and catch it. You may need to repeat the action for students who want to take a closer look as they construct their graphs.

Exchange papers with the student next to you.

2. Study the graph you just received. In one or two sentences, describe the motion of the chalk that *the graph* describes (even if that does not correspond to what you observed). Be sure to include in your description the answers to the questions:

   "How high did the chalk go?"

   "How long was the chalk in the air?"

3. Briefly discuss both graphs and both verbal descriptions with your neighbor. Decide together which graph you prefer as a solution to the original problem, or what a solution that is better than either one would look like.

4. Watch as your instructor repeats the chalk toss. Based on your previous experience, and your discussion with your neighbor, sketch a new graph.

Discuss student responses before going on. Some will not understand how to represent the passage of time on the graph, though most will see that displacement is a vertical distance (because of the particular choice of observation). The questions in 2 lead naturally to consideration of scale and coordinates. Most students will probably have the parabolic shape. If so, ask them why the top of the graph is rounded. If they have a sharp corner at the maximum, ask then what they think is happening at that time. Usually someone will say something about the chalk slowing down before it changes direction. This leads to the beginning of a discussion of velocity.

We suggest that you avoid attempting to express the relationship algebraically at this point. The emphasis in this activity is on good *graphical* representation of functions.

Watch as your instructor tosses the chalk several more times. After each toss, graph what you see, and compare your new graph with the previous ones.

Your instructor may ask you to discuss your work with another student, and synthesize ideas.

Toss the chalk again, and include some surprising variations. Here are some suggestions:

1. Toss the chalk higher or lower than the first time, and allow it to fall to the floor rather than catching it. The discussion that follows can include speculation about how fast the chalk was moving at various times, possibly leading to their drawing a graph of speed. If the discussion seems to be heading in that direction, this might a good time to suggest the idea of directed (negative) velocity, but only as a preview.

2. Toss the chalk so high that it hits the ceiling rather forcefully. After they have constructed their graphs, discuss how to represent the high point.

3. Toss the chalk against the board so that it hits the board and drops down into the chalk tray.

Classroom walk

Topic: Graphical modeling

Summary: Students observe a physical motion and record the position function as a graph.

Background assumed: The "Chalk toss" activity is helpful.

Time required: 15–20 minutes

Threads: graphical calculus, modeling, distance and velocity

Your instructor will walk across the front of the classroom. He or she will designate a "starting line" in the front of the room, so there will be a reference point for the trip.

1. You are to graph the displacement, f, of your instructor from the starting line, as a function of time, t.

This should be done for at least three "trips." The first is a simple walk in one direction. (Be sure to tell the students which direction is the positive direction.) The students need to see that this horizontal displacement from the starting line translates into a vertical distance on the graph. In the discussion, velocity can be graphed as well.

2. Graph the second trip.

The second trip should involve walking away from the starting line, turning around and walking back, with a fairly long pause somewhere during the trip. When you discuss this afterwards, be sure the students understand that the distance was maximum when you turned around. You will need to say something about negative velocity; point out that the maximum occurred when the velocity changed signs.

3. Graph the third trip.

This last walk can involve displacements in both the positive and negative directions. Many variations are possible. We refer back to this activity throughout the course to illustrate optimization, concavity, and many other topics.

Problems

1. Biking to school
2. Raising a flag

Biking to school[1]

[1] Adapted from Neil Davidson, ed., *Cooperative Learning in Mathematics: A Handbook for Teachers*, Addison-Wesley, 1990.

Topic: Graphical modeling

Summary: Students match stories with graphical representations of distance functions.

Background assumed: "Chalk toss" and "Classroom walk" are helpful.

Time required: 5 minutes

Threads: graphical calculus, modeling, distance and velocity

Terry usually rides a bicycle to school. Below are four graphs and three explanations. Match each explanation with a graph, and write an explanation for the remaining graph.

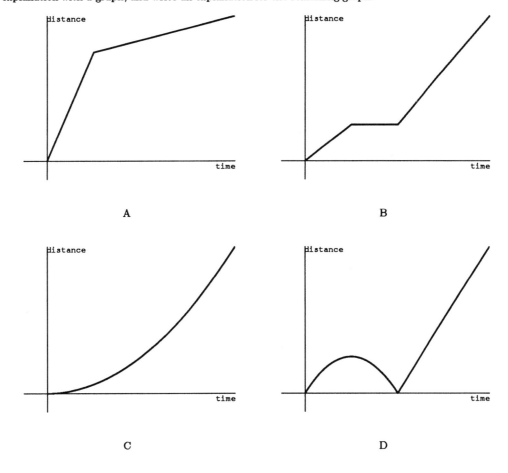

1. "I had just left home when I realized we have gym today, and I had forgotten my gym clothes. So I went back home and then I had to hurry to be on time."

2. "I always start off very calmly. After a while I speed up, because I don't like to be late."

3. "I went on my motor bike this morning, very quickly. After a while, I ran out of gas. I had to walk the rest of the way and was just on time."

4.

Students often notice that the sharp corners on the graphs are not realistic representations of motion. An informal discussion of why—discontinuous velocity—is fine here.

Raising a flag

Topic: Graphical modeling

Summary: Students represent a function as a graph from imagining or observing the raising of a flag.

Time required: 5 minutes

Threads: graphical calculus, modeling

---

1. Sketch the height of a flag as a function of time as it is being raised.
2. Sketch the height of a flag as a function of time as it is being lowered.

---

1. Get their ideas and discuss them.

2. Show alternative solutions and discuss these. The two most common ones are a straight line with positive slope (the flag goes up at a constant rate), and a non-decreasing function which alternates horizontal pieces and strictly increasing pieces (the flag is raised hand over hand, with brief pauses).

3. Discuss the effect of the vague problem description on the resulting proposed solutions. (We show a video of a flag being raised and lowered, and then we ask which of the proposed solutions best describes what they saw in the video.)

Library trip

Topics: Graphical modeling, velocity from distance graph

Summary: Students construct and interpret graphs of distance functions, given verbal descriptions.

Background assumed: "Chalk toss" and "Classroom walk" are helpful.

Time required: 20 minutes

Threads: graphical calculus, modeling, distance and velocity

In this activity we will describe a situation verbally and ask you to try to construct a graph that might correspond to it.

> Josh had arrived a little early for his calculus class when he realized that he had left his notebook in the library. Not wanting to miss the beginning of class, he hurried directly to the library, picked up his notebook, and returned just as quickly to the classroom. The library is 500 meters directly across the quad from the classroom, so you can assume he walked back and forth in a straight line. The entire trip took six minutes.

1. Construct a graph describing Josh's trip to the library and back. The independent variable is time, t, and the dependent variable is the distance, f, between Josh and the classroom.

Be sure the students have used an appropriate scale on both time and distance axes. Notice also that the phrase "just as quickly" tells something about the slope of the graph on two different intervals. Although they may not yet have thought about that connection, this is a natural place to begin discussing it, especially if some students bring it up.

---

2. Now let's embellish the story a little:

> On the way to the library Josh met Pat and stopped to talk for three minutes. Then he had to move even faster for the rest of the trip. It took him a total of eight minutes to make the entire trip.

As before, record the trip on a graph.

3. Now that we have seen how motion can be recorded on a graph, let's look at some of the graphs you have drawn, and see how important information is shown on the graph.

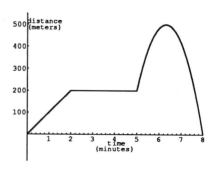

Let's look first at a typical graph for question 2—Josh's interrupted trip to the library.

Library trip

Label on the graph the locations of the following (label the graph with the letter of the corresponding question):

(a) The period of time that corresponds to Josh's conversation with Pat.

(b) The time or period of time that corresponds to Josh's presence in the library.

(c) A period of time when the distance between Josh and the classroom is increasing.

(d) A period of time when the distance between Josh and the classroom is decreasing.

(e) A time or period of time when the distance between Josh and the classroom is a maximum.

(f) A time or period of time when Josh is not moving at all.

An important concept associated with motion is that of velocity. Velocity represents both the speed at which something is moving and the direction in which it is moving. For this to make any sense, we have to agree in advance about what we will regard as the "positive direction." The opposite direction is then the "negative direction."

For this problem, suppose we agree that an arrow pointing from the classroom toward the library is pointing in the positive direction, and an arrow pointing in the opposite direction is pointing in the negative direction. Thus, when Josh is moving away from the classroom, his velocity is positive, and when he is moving toward the classroom, his velocity is negative. Aside from these considerations, velocity is just speed. That is, Josh's speed is the absolute value of his velocity.

4. Now try to label these additional areas on the graph:

(a) A period of time when the velocity is positive.

(b) A period of time when the velocity is negative.

(c) All the times when the velocity is zero.

(d) Two times or periods of time, $4d_1$ and $4d_2$, when the velocity is positive, but chosen so that the velocity at $4d_2$ is greater than the velocity at $4d_1$.

(e) A time or period of time when the velocity is positive, but Josh is slowing down.

Follow up:

Have the students sketch a graph of the velocity from the information in question 4, and the graph in question 3.

Airplane flight with constant velocity

Topics: Distance-velocity relationship for constant velocity

Summary: Students work through an example of constant velocity and linear distance to see that velocity is the slope of the distance function and that distance is the area under the velocity graph.

Time required: 50 minutes

Threads: graphical calculus, modeling, distance and velocity

Part A

A plane flies over Ithaca on its way to Los Angeles at 1 o'clock in the afternoon. The plane's velocity is 500 mph, and for the next four hours the plane flies at a constant velocity of 500 mph. The plane flies at the same altitude for the entire time.

1. Sketch a graph of the velocity of the plane versus time.

After or while the students draw the graph, be sure to talk about the idea of scale. For example, does the velocity scale need a tick at 2000? Which axis is vertical and which is horizontal? Note that time should go from 1 to 5 (or time should be defined as time after 1 o'clock). If they include values outside that domain, you can discuss whether they know these and whether they are plausible.

The graph should look like this.

Constant velocity

Since the velocity at 3:30 is 500 mph, your velocity graph should have a point with coordinates $(3.5, 500)$. Remember the first coordinate in the coordinate pair indicates the time. Since you were not told what the velocity is at 12:15, there is no point on the graph whose first coordinate is 12.25.

2. Sketch a graph of the *distance* of the plane from the place where it passed over Ithaca versus time.

3. Compute a table of values of the distance at 1, at 2, at 3, at 4 and at 5 o'clock to check your graph in 2. If you were not able to sketch the graph, use the table of values to locate points on the graph and then go back and sketch the graph.

Time	Distance
1	
2	
3	
4	
5	

4. To make sure you understand this example draw velocity graphs of a plane that had a velocity of 600 mph and a plane that traveled at a velocity of 400 mph. Then draw the distance graphs for these planes on the same set of axes as you drew the distance graph of the original plane.

5. Look at the distance graphs and try to relate the distance graphs to the velocity graphs.

The students should observe that the distance graph is a straight line when the velocity is constant. Almost everyone will say that the "higher" or steeper distance graph corresponds to a faster speed. Many will notice that slope corresponds to velocity.

When you learned about straight lines you learned the concept of the slope of a line. If you are graphing a variable y on the vertical axis and a variable x on the horizontal axis and (x_1, y_1) and (x_2, y_2) are the coordinates of two points, then the *slope* of the line segment between the two points is $(y_2 - y_1)/(x_2 - x_1)$.

Another very important way to think of the slope is that it gives the *rate of change of the quantity measured on the vertical axis with respect to the quantity measured on the horizontal axis.* Let's work through that phrase in detail for the example we just worked out to make sure we understand what the phrase means. We plotted distance from where the plane passed over Ithaca on the vertical axis and time on the horizontal axis. Two points on the distance graph were $(3, 1000)$ and $(5, 2000)$. Computing the slope between these two points gives you $1000/2 = 500$.

Keep in mind that the vertical axis is measured in miles and the horizontal axis in hours, so we should interpret the slope as a way of telling us that the distance must increase by 1000 miles when the time increases by 2 hours or by 500 miles whenever the time increases by 1 hour. This is written as 500 miles per hour, which is exactly the velocity of the plane.

So we can think of *velocity* as *the rate of change of distance with respect to time.* This is indicated by the units used to measure velocity such as miles per hour, centimeters per second, etc.

Geometrically, a slope of 500 means that to stay on the graph we move up 500 units each time we move 1 unit to the right.

6. Find the slope of the distance graph if the plane had a constant velocity of 400 miles per hour. Be sure to include units of measure.

7. Based on this activity, we have seen that *if the distance graph is a straight line then the* _____ *of the straight line is the velocity.*

Part B

After the class has done Part A, it's a good idea to do Part B in class as a teacher guided activity and discussion. Then, have them repeat the process and verify the conclusions for a velocity of 600 miles per hour.

At the start of Part A, you were able to draw the graph of the distance traveled by looking at the velocity graph and reasoning about the physical quantities involved. Let's look for a way to use this same reasoning to represent distance traveled *directly on the velocity graph*.

1. Consider any time period, say for example the first half hour (from 1:00 to 1:30). How do you calculate the total distance traveled? It is the velocity multiplied by the elapsed time. That is, $d = v \times t$. For the plane traveling at 500 miles per hour, calculate the distance traveled:

    (a) in the first half-hour.
    (b) in the first hour.
    (c) in the first two hours.
    (d) in the first three hours.

In this example, the velocity graph is a horizontal line, and the magnitude of the velocity at any time is the vertical length measured from the t-axis to the velocity graph, always 500 mph in this case. The elapsed time is a length measured along the horizontal axis. In this case it is the length of the part of the horizontal axis between $t = 1$ and $t = 1.5$, that is, 0.5.

So the distance traveled, 500 mph × 0.5 hr, or 250 miles (see 1a) is also the *area* of the rectangle shaded at the right.

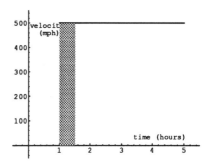

First half hour

Note the units. Miles per hour times hours yields miles. (Remind the students that "per" means "divided by.")

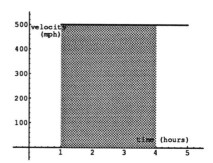

First three hours

We could calculate the distance traveled in any other time period in a similar way. In each case, the distance traveled is also the area of a corresponding rectangle.

2. Calculate the area of the rectangle above. Compare your answer with 1d.

3. Draw a rectangle whose area corresponds to:
   (a) the distance you computed in 1b.
   (b) the distance you computed in 1c.

4. For this example (constant velocity and linear distance), at least, we seem to have found the following pair of principles:

   If you have a graph of distance traveled, velocity at some specific time is the _____ of the distance graph (at the appropriate point).

   If you have a graph of the velocity, distance traveled during some time period is the _____ _____ the velocity graph (over the appropriate interval).

We have also discovered another relationship. We have found a correspondence or pairing of two graphs. Namely:

If you have a horizontal velocity graph, then the distance graph is linear.

If the distance graph is linear, then the velocity graph is horizontal.

This idea of pairs of graphs will reappear throughout the course. Finally, to finish summarizing the general principles we have developed:

Velocity is the rate of change of distance with respect to time.

Students can work out other examples if these ideas are not clear at this point.

In real life (even with cruise control) it would be difficult to fly a plane or drive a car at *exactly* the same speed for a long period of time. However we will use simple examples such as the case of constant velocity to analyze more realistic and more complicated problems. The idea of using simple problems to attack complicated problems is a fundamental strategy in mathematics.

Problems

1. For the plane in the preceding section that traveled at a constant velocity of 500 mph, express the velocity, v, as a function of time, t. That is, write a formula for v in terms of t.

   Also express the distance traveled, f, as a function of t. That is, find $f(t)$.

   Observe how the principle "velocity = slope of distance" looks in terms of the formulas.

   Repeat the problem for the other two velocities, 400 mph and 600 mph.

2. Suppose the velocity is 500 mph from 1 to 3 and 400 mph from 3 to 5. Construct the distance graph using the slope-velocity relationship (your graph will look like two line segments joined together), and verify that distance = area.

This is more difficult. The students may worry about what is happening at time 3.

Projected image

Topics: Modeling, curve fitting

Summary: Students collect data, construct models to fit the data, and use the models to make predictions.

Time required: 50 minutes

Resources needed: Overhead projector on a wheeled cart.

Thread: modeling

Part A

1. Watch as your instructor moves the overhead projector. Describe in your own words what happens.

Turn the overhead projector on, with its image projected on a wall. Move the projector further from the wall and/or closer so the size of the image changes. Expect the students to describe how the size of the image is changing and to observe that the size depends on the distance of the projector from the wall. There will be various ideas about how one measures image size. That is fine, and each student or group should be allowed to use its own definition.

---

2. Now collect some experimental data. Measure the distance of the projector from the wall and the image size at three different points. Record your experimental data in the table below.

Distance from wall	Size of image

The students will experience first-hand the problems of measurement and approximation. The class can do the data collection as a single group to save time.

---

3. Now, as a group, sketch what you believe is a reasonable graph of size of image versus distance from the wall. Be sure to label your axes clearly and include a scale and units.

4. Write a verbal description of what your graph says happened as the projector was moved.

5. Compare the verbal description you just wrote with the description you wrote before collecting the data and drawing the graph.

6. Now choose a distance that you did not measure experimentally, and predict what the image size would be at that distance. Measure and check your prediction.

Check some of the predictions and calculate errors. Discuss the possible sources of error (measurement, prediction).

Part B

Below is the experimental data collected in one of our classrooms. As with the data you collected, some of our measurements may be in error. We decided to measure "size" as the height of the projected image. We have transferred the points to a graph.

Distance from wall	Size of image
18	13
24	17
48	29.5

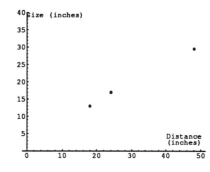

Projector data

As you probably observed when you made your own measurements, errors in making the measurements have almost certainly occurred, so even the little data we have is not necessarily accurate. And the three points on the graph, of course, give a very incomplete picture of the relationship between distance and image size. For instance, how big is the image when the projector is 42 inches from the wall? Or at any other distance, x, from the wall?

1. Based on the three data points given above, predict the size of the image when the projector is 42 inches from the wall.

2. If we have only this experimental data, what kinds of graphs are reasonable models based on the data? *The key question to ask is: what happens between the data points?* There are lots of ways to fill in the graph. Several possibilities are given on the next page. For each graph given:

    (a) What size image does the graph predict when the projector is 42 inches from the wall?

    (b) Describe in your own words why it is or is not a reasonable graph to use to represent the data and the experiment.

    (c) Choose the graph that you think best describes the size of the image as a function of the distance of the projector from the wall.

3. For as many of the graphs as you can, find a formula for the size of the image in terms of the distance from the wall.

4. Suppose the projector is 24 inches from the wall. For each graph, predict what would happen if the projector is moved a short distance

    (a) closer to the wall.

(b) further from the wall.

The class probably decided that the "jumps" in the first graph (A) are not very plausible, although as the course goes on they will see that graphs like it are appropriate for some situations, and that such "jumps" in a graph are called *discontinuities*.

Most people would also agree that the last graph (E) is a poor representation of the situation, even though it goes right through the three data points, because it behaves so erratically in between.

On the other hand, the other three graphs aren't too bad. You may want to discuss each of them in more detail.

The second graph (B) is a *piecewise-linear* graph. It consists of two line segments "pasted together." Fitting a piecewise linear graph to the data has the advantage that the graph has a formula that is easy to write down. In this case, the formula is:

$$f(x) = \begin{cases} \frac{2}{3}x + 1, & x \leq 24 \\ \frac{25}{48}x + \frac{9}{2}, & x \geq 24 \end{cases}$$

Point out that this formula consists of two expressions. The first expression is used for some of the possible values of x and the second is used for other values of x. When $x = 24$ either expression can be used. In this formula that is not a problem, since both expressions give the same value when $x = 24$.

We can use the formula for making predictions. Our formula predicts a size of $\frac{211}{8}$ or about 26.5 inches at a distance of 42 inches from the wall. To see this, substitute 42 inches in the appropriate expression (in this case, it is the second expression, since 42 is between 24 and 48). This gives a value of $\frac{25}{48}(42) + \frac{9}{2}$, which reduces to $\frac{211}{8} = 26.5$ inches. Predicting values between data points based on a piecewise linear graph is called "linear interpolation."

The difficulty with the second graph is that it isn't smooth. According to this piecewise linear graph, the rate at which the image size changes is abruptly different when you switch from one linear piece to the other. Our intuition should tell us that the piecewise linear graph, while useful for some purposes, cannot be a completely accurate model of this physical situation.

The third graph (C) is a single line chosen in such a way as to fall as close as possible to the whole collection of data points—the *least squares line*. The formula for the line in this graph is $y = \frac{13}{24}x + \frac{43}{12}$. It therefore predicts a size of $\frac{79}{3}$ or about 26.3 inches at a distance of 42 inches from the wall.

The difficulty with this graph is that none of the data points actually lie on the graph, even though all are close. It would make a prediction different from the experimental evidence even at the distances for which we have measurements! (Of course, as we discussed earlier, our measurments–the data–may contain errors, so they are not completely reliable.)

The fourth graph (D) was drawn freehand, in such a way as to hit all the data points and represent a pretty smooth curve. In drawing it, we also tried to avoid all those wild fluctuations you saw on the fifth graph. We make predictions from the fourth graph by graphical estimation. If we use the fourth graph, it looks like the image size at a distance of 42 inches would be approximately 27 inches.

The difficulty with the fourth graph is that there are many freehand curves that could be drawn through the data points. In fact, if two people each drew a freehand curve through these three points, their graphs would almost certainly be at least a slightly different, although they should probably have the same general shape. So while a freehand curve gives a good overall view of the ups and downs, bends and so forth, specific predictions based on such a curve are not fully reliable. Besides, we have no reason to believe it represents the true picture, any more than the linear or piecewise linear solutions.

The point is that there are many ways to reconcile the desire to make predictions with a scarcity of data points. Some of these methods are purely graphical, but others result in formulas. There are procedures for finding formulas for smooth curves that fit data as well. The result of using a given procedure can be a nice curve like the fourth graph, but it might also look a lot like the fifth one. The lesson is that there are lots of possibilities and lots of reasons to be cautious. Not the least important reason to be skeptical is the idea that no matter how "good" a graph looks, the ultimate question is, "How well does it model the real world problem with which we started?"

Problems

1. What does each graph above tell you about the size of the image when the projector is 36 inches from the wall?
2. Describe in a few sentences what each graph says about how the size of the image changes as the projector is moved from the wall.

A formula for a piecewise-linear graph: Top-down analysis

Summary: The derivation in this activity has two purposes. It introduces the important *top-down* approach to problem solving. This is crucial to the emphasis on mathematical modeling that we take in the course. It also introduces the idea of generalization, by solving a problem with parameters, and applying the solution to the particular values at hand.

Time required: 20 minutes

Thread: top-down methodology

The second graph from the projected image activity is shown at the right.

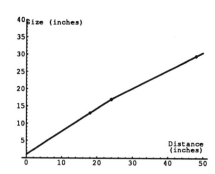

Projector image

This is a piecewise-linear graph; that is, it is obtained by "pasting together" line segments. In this section we will discuss how to get a formula for this (and any other) piecewise-linear graph. Let us look in some detail at how this is done.

Our solution to this problem uses a top-down approach. That is, we will determine what big steps are needed to solve the problem, then reduce each big step to smaller steps. We repeat this process, breaking each smaller step into even smaller ones until we recognize the smallest pieces as problems we know how to solve. We then reassemble the solutions to the small pieces successively until we have a solution to the original large problem.

Let's see how this works for the problem of writing a formula for the piecewise-linear graph.

In presenting the following material in class, it is a good idea to construct the outline form first—including some discussion of motivation—and then return to fill in the solutions and do the reassembling process. The outline form is at the end of the section.

Level I (top level)

The formula we want describes two line segments; call them L_1 and L_2. The formula for line L_1 applies only between the first and second data points, and the formula for L_2 applies only between the second and third data points. So we can find the final answer if we can find the answers to two smaller problems:

I.A. Find a formula for line L_1 when $x \leq 24$, and
I.B. Find a formula for line L_2 when $x \geq 24$.

So the first step in the top-down analysis of the problem (and the last step we do when we are reassembling the pieces) is:

I. Write the formula for the piecewise-linear graph as:

$$y = \begin{cases} \text{formula for line } L_1 & \text{when } x \leq 24 \\ \text{formula for line } L_2 & \text{when } x \geq 24. \end{cases}$$

Level II (smaller problems we need to solve in order to complete level I)

It should be pretty clear that in order to complete I, we need to be able to do two things:
Write the formula for L_1, a line that goes through the two points $(18, 13)$ and $(24, 17)$.

Write the formula for L_2, a line that goes through the two points $(24, 17)$ and $(48, 29.5)$.

So if we can solve these two problems, then we can solve the original problem. This leads us to level III.

Level III (smaller problems we need in order to complete level II)

In order to find the equation of a straight line when we are given two points we can use the point-slope form of the equation for a straight line, which is

$$y - y_1 = m(x - x_1).$$

In this formula, m is the slope of the line and (x_1, y_1) are the coordinates of a point on the line. So in order to solve I.A we need to solve three smaller problems:

I.A.1. Find the slope of the line through $(18, 13)$ and $(24, 17)$.

I.A.2. Use one of the two given points and the point-slope formula to find the equation of the line through $(18, 13)$ and $(24, 17)$.

I.A.3. Solve the equation that we found in I.A.2 for y to get the formula for L_1.

Similarly, in order to find the equation for L_2 we will need to solve the same three problems.

Now we will solve these problems and assemble a solution to the original problem.

First we will solve I.A.1.: the slope is $\frac{17-13}{24-18} = \frac{2}{3}$. Using $(18, 13)$ as the given point gives $y - 13 = \frac{2}{3}(x - 18)$ as the result of I.A.2. Finally, solving for y gives us $y = \frac{2}{3}x - 12 + 13 = \frac{2}{3}x + 1$ as the outcome of I.A.3.

Similarly, the slope is $\frac{29.5-17}{48-24} = \frac{25}{48}$. Using $(48, 29.5)$ as the given point gives $y - 29.5 = \frac{25}{48}(x - 48)$. Finally, solving for y gives us $y = \frac{25}{48}x - 25 + 29.5 = \frac{25}{48}x + \frac{9}{2}$.

Solving the level III problems has given us the solution to the level II problems.

The equation of line L_1 is $y = \frac{17-13}{24-18}(x - 18) + 13$ or $y = \frac{2}{3}x + 1$ and the equation of line L_2 is $y = \frac{29.5-17}{48-24}(x - 24) + 17$ or $y = \frac{25}{48}x + \frac{9}{2}$.

These two equations are the solutions to the two subproblems at level II.

Returning finally to level I of the original problem (reassembling):

The solution is: the formula for the piecewise-linear graph is

$$f(x) = \begin{cases} \frac{2}{3}x + 1, & x \leq 24 \\ \frac{25}{48}x + \frac{9}{2}, & x \geq 24. \end{cases}$$

We can often summarize the top-down analysis of a problem in outline form. The outline looks a bit like one you would use in organizing an English paper. The outline should really be developed in advance and be used as a guide in the solution of the problem. We have reversed the procedure here to help you appreciate the thinking that goes on in designing the solution.

Here is the outline form of the top-down structure of our solutions.

Problem: Write a formula for the piecewise-linear graph.

I Write the formula for the piecewise-linear graph as

$$y = \begin{cases} \text{formula for line } L_1 & \text{when } x \leq 24 \\ \text{formula for line } L_2 & \text{when } x \geq 24. \end{cases}$$

A Write the formula for L_1, a line that goes through the two points $(18, 13)$ and $(24, 17)$.

1. Find the slope between $(18, 13)$ and $(24, 17)$.
2. Find the point-slope equation of the line between $(18, 13)$ and $(24, 17)$.
3. Solve for y to get the formula for L_1.

B Write the formula for L_2, a line that goes through the two points $(24, 17)$ and $(48, 29.5)$.

1. Find the slope between $(48, 29.5)$ and $(24, 17)$.
2. Find the point-slope equation of the line between $(48, 29.5)$ and $(24, 17)$.
3. Solve for y to get the formula for L_2.

Problems

1. Fit a piecewise-linear graph to the three experimental data points you collected in class. Determine the corresponding formula for the graph.

2. Observe the length of the cafeteria line in the dining hall at three different times. Record line length versus time in a table. Fit a piecewise-linear graph to the data. Include both a sketch of the graph and its formula.

Suggested homework followup: Discuss the results from problem two. If students took all their measurements during peak hours, their data will indicate a fairly flat graph. The instructor should ask them to predict the length of the line, say, just before the doors close. Most will indicate that the line should be much shorter at that time. Ask them to predict the length of the line half an hour after the doors close. Point out that they are basing the last two conclusions on theoretical considerations about the behavior of the system, not on the data they collected. (This is a good preview of the next activity ("Water balloon"), in which modeling from theoretical considerations about the problem is covered.) Also, point out the difference between interpolating between data points and extrapolating outside the domain of the collected data.

Water balloon

Topic: Modeling

Summary: A theoretical model is derived and compared with actual data.

Background assumed: Formula for the volume of a sphere.

Time required: 25–30 minutes

Resources needed: Measuring beaker, balloons

Threads: modeling, multiple representation of functions, top-down methodology

---

This is another physical experiment in which we will compare observed data with a mathematical model. Your instructor has a balloon filled with water. The water will be released into a beaker a little at a time. At several points in the process you will measure the circumference of the balloon, and also observe how much water is in the beaker. That is, you will be observing the relationship between the balloon's circumference and the amount of water that has been released.

Before the class actually does this experiment, they should discuss the units that they will use. This will depend on the equipment that the instructor has been able to provide. Lacking a suitable laboratory, it will probably be most convenient to collect the data as a class (using volunteers) and to do the analysis in small groups.

Record your experimental data in the table below.

Circumference in _____ (units)	Amount of water released in _____ (units)

To predict the volume of water released at circumferences between those you observed you could fit a piecewise-linear or other graph to the data. In fact, you will probably be asked to do that as a homework problem.

Another approach is to *make a mathematical model in the form of a formula directly from the (theoretical) geometric properties of the balloon and the stated description of the problem.*

Before you begin, let us agree to call the circumference of the balloon C, and the amount of water released A. The formula you derive, therefore, will be of the form

$A =$ some formula whose variable is named C.

In fact, to emphasize that the value of A will depend on the value of C, we will write $A(C)$ in place of A. That is,

$A(C) =$ some formula whose variable is named C.

Find a formula for $A(C)$.

The class should work on this problem in their groups for at least ten minutes (with the instructor observing and making suggestions). They will have to decide how to model the shape of the balloon. The most obvious decision is to assume the balloon is spherical. They may need to be told the formula for the volume of a sphere. They also need to decide whether or not to use the measured initial circumference of the balloon as a constant in the formula. If any groups show an inclination to generalize this to a parameter they should, of course, be encouraged to do so.

After the groups have had a sufficient chance to work on this problem, the solution should be discussed with the whole class. This is a good place to use the top-down approach. This problem breaks down as:

1. Compute volume of water released as:

    Volume released = Initial Volume − Volume Remaining .

    (a) Compute initial volume as a function of initial circumference, C_0.

i. Use the formulas $V = 4\pi r^3/3$ and $C = \pi r^2$
ii. Solve the formula for r in terms of C, and substitute C_0 for C.
iii. Substitute the solution to ii. in i.

(b) Compute volume remaining as a function of circumference, C.
  i. Use the formulas above.
  ii. Solve the formula for r in terms of C.
  iii. Substitute the solution to ii. in i.

Reassemble, by substituting the solutions to a. and b. into the "formula" in 1.

Now that we have a formula for the amount of water released, based on geometric considerations, we should compare the values predicted by our formula with those we obtained by actual measurement. That is, compute $A(C)$ for all the values of C that you measured, and compare the values generated by the formula with the amounts of water you measured.

This process should lead to a discussion of two important sources of error: measurement error and modeling error introduced by simplification—we idealized the shape of the balloon in order to use to a familiar geometric formula. Most students will believe the measurements are more accurate than the values computed from the derived formula.

Problems

1. Fit a piecewise-linear function to the experimental data points you collected in class. Determine the corresponding formula for the graph. If you have graphing software available, compare the graph of the formula you derived in class with the piecewise-linear one above.

2. Derive a formula for the projected image model, from theory. (See the "Projected image" activity.)

Graphical estimation of slope

Topic: Slopes of non-linear graphs

Summary: Uses software or graphing calculator to approximate a curve by a straight line. The slope of this line is an approximation to the derivative of the original curve.

Background assumed: Some experience with the software or calculator

Time required: 30 minutes

Resources needed: Graphing calculator or computer software capable of "zooming in" on a graph.

Threads: approximation and estimation, graphical calculus

This activity assumes the availability of graphing calculators, or of computer software with the capability of "zooming-in" on graphs. If this technology is not available to students, the instructor could present the materials as demonstrations, or simply make overhead transparencies that zoom in on a graph by repeatedly enlarging sections of the graph, using the enlarging capacity of a duplicating machine.

In this activity, we will examine the concept of "slope" for graphs that are not straight lines. The idea is the following: if you consider only a very small portion of a curve, that curve might appear to be a straight line. As an analogy, assuming that the earth is a sphere (it's not, but it's close), you could walk along the equator and feel that you were walking along a straight line, even though an observer in space would realize that the equator is curved.

The same is true of certain non-linear graphs or curves. Select a point on the graph on which to focus your attention. If you look at a small enough piece of the graph near your point, it looks like a straight line. That apparent straight line has a slope. This slope is a good approximation of what we call the slope of the graph at the point where you are focusing attention. (Some graphs never start to look straight no matter how closely you look. We'll discuss that problem in the second part of the activity.)

Part A

1. We will begin to examine these ideas by looking at the function $f(x) = 10(x^2 - x^3)$, near the point $(0.5, 1.25)$. First, you should verify that the point $(0.5, 1.25)$ lies on the graph of f. (How?)

2. Now, using graphing software on a computer or a graphing calculator, graph f. Be sure that your view includes the point $(0.5, 1.25)$ and is large enough to give you a good overall view of the shape of the entire graph.

Now we want to get closer and closer views of this graph, near the point $(0.5, 1.25)$.

Your instructor will tell you how to "zoom in" on a point on the graph, using your software or calculator. In some cases this may be just a matter of selecting a small rectangle on the display, and "blowing up" the picture. In others, you may need to select the x and y limits of the display explicitly.

3. Blow up the display, using a rectangle from about $(0, 0.75)$ (lower left) to $(1, 1.75)$ (upper right). Does the graph appear to be a straight line? If not, blow up the display again. This time use a rectangle from about $(0.4, 1.15)$ to about $(0.6, 1.35)$. Keep zooming in until the graph seems to be a straight line.

4. Once you are satisfied that the graph looks straight from close up, estimate its slope from the display. This requires reading the coordinates of two points from the screen, since we need two points to determine slope. Fortunately, we know one point that is displayed on this part of the graph—it is the point on which we have been focusing attention, $(0.5, 1.25)$.

   Now we need to find a second point on the graph. Your software may have a feature thats helps you find the coordinates of a point on a display. If not, you will need to estimate it as best you can. Record the (approximate) coordinates of the second point.

Use these two point to calculate the approximate slope of the line.

When you finish, redisplay the graph, returning to the original domain and range, and continue.

5. Zoom in once again on the graph of $f(x)$, this time focusing on the point $(0,0)$. Find two points and the slope.

6. The second point we used in computing the slope of the "line" is just a visual estimate. That is, we cannot be sure that it is actually a point on the "line." How can you exploit the fact that this apparent "line" is actually a portion of the original graph in order to get exact coordinates of a second point?

They should decide that they can get an actual point on the graph by substituting their estimated x-value into the expression for $f(x)$. They are computing the slope of a secant line.

Part B

1. Now that you know how to estimate the slope of a function graphically, you are to obtain such estimates for several points on the graph of $g(x) = 2x - 3x^2$.

   Fill in the table at the right.

Focus point	Approximate slope
$(-2, -16)$	
$(-1.5, -9.75)$	
$(-1, 5)$	
$(-0.5, -1.75)$	
$(0, 0)$	
$(0.5, 0.25)$	
$(1, -1)$	
$(1.5, -3.75)$	
$(2, -8)$	

2. When you have completed your table, observe whether or not there seems to be any pattern. Can you guess a way of determining the slope at the point $(1.1, -1.43)$ without using the computer? How about at the point $(a, 2a - 3a^2)$?

If they have completed the table accurately, and looked for a pattern, they will predict a slope of -4.6, $(2 - 3(1.1))$, at $(1.1, -1.43)$ and a slope of $2 - 6a$ at $(a, 2a - 3a^2)$.

Part C

1. See what happens for the function $h(x) = |x|$.

   Fill in the table at the right.

Focus point	Approximate slope
(-1, 1)	
(-0.5, 0.5)	
(0, 0)	
(0.5, 0.5)	
(1, 1)	

2. When you have completed your table, observe whether or not there seems to be any pattern. Pay special attention to $(0,0)$. How about at the point (a, a) for $a > 0$? How about at $(a, -a)$ for $a < 0$?

The important point here is the discontinuity in the slope function at the point $(0,0)$ on the original graph. This is a good time for them to articulate for themselves what property of the graph of $h(x) = |x|$ corresponds to the presence of a discontinuity in the slope function.

Here are the graphs we got for part A.

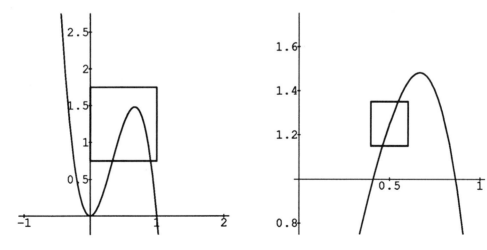

Initial view   Second view

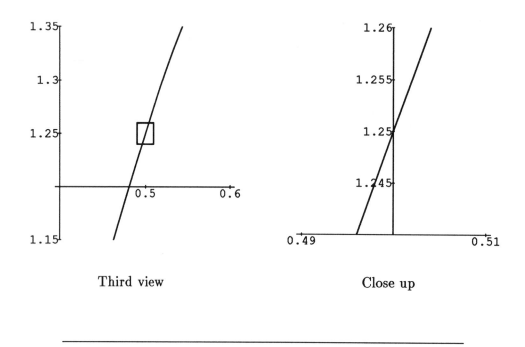

Third view   Close up

Slope with rulers

Topics: Slopes and derivatives

Summary: Students calculate slopes from graphs, by using a straight edge.

Time required: 5 minutes for each graph

Resources needed: Straight edge

Threads: graphical calculus, multiple representation of functions

For each of the four graphs below, estimate the slope at a number of points. Can you find a function that gives the slope as a function of x?

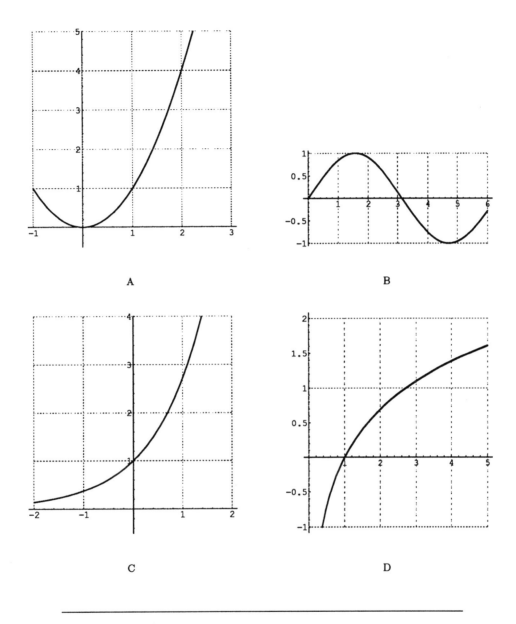

A: $f(x) = x^2$. (They should know this one. Find slopes at $x = 0, 1$, and 2 to verify what you know.)

B: $f(x) = \sin(x)$.

C: $f(x) = e^x$. Estimate slopes at $x = -1, 0$, and 1. See if they can guess the derivative.

D: $f(x) = \ln x$. Have them estimate slopes at $x = 1, 2, 3$, and 4. See if they can guess the derivative.

Some helpful ideas for computing the slope with a ruler are:

1. Use two points that are far from each other rather than two points that are close to each other in order to minimize the error introduced by reading coordinates from the graph.

2. Use points which are on the grid if at all possible.

Examining linear velocity

Topics: Finding distance for a linear velocity function

Summary: Students find the distance function for a linear velocity, and relate the distance to the area under the velocity curve. They compute the slope of the distance function and relate this slope to the velocity.

Background assumed: "Airplane flight with constant velocity"

Time required: 20–25 minutes

Threads: graphical calculus, distance and velocity, approximation and estimation

---

Returning to the relationship between distance and velocity, suppose we have a vehicle that starts at rest (velocity = 0) and speeds up at a constant rate. (By that we mean that the vehicle speeds up in such a way that its velocity graph is linear.) Let's assume the velocity has the formula $v(t) = 2t$, for $0 \leq t \leq 8$. Assume that time is measured in seconds and distance in feet, so velocity is measured in feet per second. Then the velocity graph looks like this.

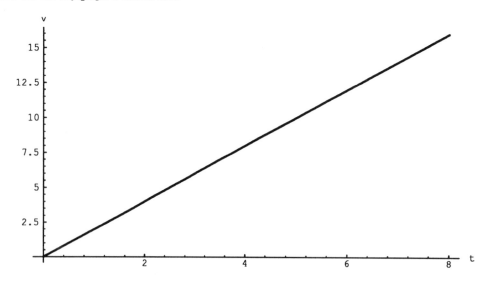

Constant velocity

Can we sketch a reasonable graph of the distance traveled?
Do the two principles

$$\text{distance traveled} = \text{area under the velocity graph}$$
$$\text{and}$$
$$\text{velocity} = \text{slope of the distance graph}$$

still seem to hold?

Part A

1. Finding out exactly what the distance graph looks like is no longer such an easy problem. One way we can approach it is by computing some approximations. During the first second, the car speeds up from 0 feet per second (fps) to 2 fps. So we might realistically say that its average speed during the first second is 1 fps. A reasonable approximation of how far it travels in the first second is obtained by pretending it travels at 1 fps for the entire second. The distance, therefore, is approximately 1 fps × 1 sec, or 1 foot.

   Similarly, the average velocity for the second second might reasonably be said to be 3 fps, since the speed at the beginning of the second second is 2 fps and at the end of the second second it is 4 fps. Thus, the approximate distance traveled in the second second is 3 fps × 1 sec, or 3 feet. Fill in the rest of the table at the right.

Second	Computation
1	1 fps × 1 sec = 1 ft
2	3 fps × 1 sec = 3 ft
3	5 fps × 1 sec = 5 ft
4	
5	
6	
7	
8	

2. The approximate total distance traveled in the first x seconds, using $x = 1, 2, 3, 4$ is tabulated at the right. Fill in the rest of the table.

x	Approximate total distance traveled in first x seconds	
1	1	
2	4	(i.e., $1+3$)
3	9	(i.e., $1+3+5$)
4	16	$(1+3+5+7)$
5		
6		
7		
8		

3. Now, we can get a rough sketch of the distance graph by sketching a curve through the points $(0,0)$, $(1,1)$, $(2,4)$, ..., $(8,64)$. The actual points we obtained, however, should give you a good idea of just what curve will represent the exact distance graph! What curve will represent the distance?

   They should all guess $f(t) = t^2$.

   ---

4. Now let's see how these calculations are related to area. Look again at the velocity graph. We will compute the area under the graph from $t = 0$ to $t = 1$, from $t = 0$ to $t = 2$, ..., and from $t = 0$ to $t = 8$ first using an approximating technique. Look at the diagram below. For the first second, the area under the graph (which is actually the area of a triangle) can be seen to be equal to that of the sketched rectangle. Similarly for the area for the second second, and so on.

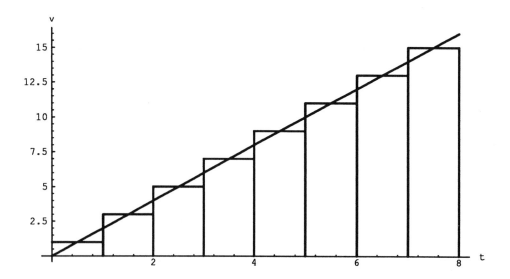

Approximate distance each second

Calculate these areas, along with the total area from $t = 0$ up to and including the current rectangle, in the table at the right.

x	Total area to the left of x
1	1
2	4 (i.e., $1 + 3$)
3	9 (i.e., $1 + 3 + 5$)
4	
5	
6	
7	
8	

Not only are the values for total area the same as those for our approximation of total distance traveled, the computations are the same too.

Once again, the relationship "distance traveled = area under velocity curve" is seen to hold. Furthermore, we seem to have another pair of related graphs, namely:

If the velocity graph is linear, then the distance graph is quadratic.

Part B

What about the second relationship: velocity = slope of distance graph? In the current problem we have a distance graph whose formula is $f(t) = t^2$.

A carefully drawn graph of $f(t) = t^2$ is given here.

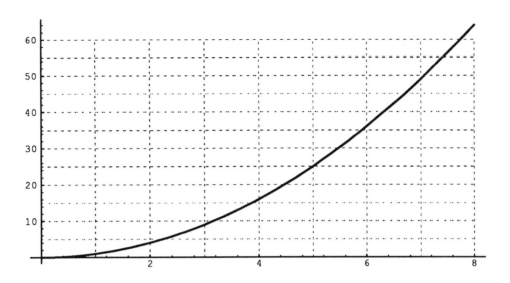

Distance traveled

1. Use the graph and a ruler to estimate the slope of the graph (i.e., the slope of the tangent line to the graph) at each of the values of t that are listed in the table at the right.

t	Slope of the graph at t
1	
2	
3	
4	
5	
6	
7	
8	

2. Guess a formula for slope, as a function of t.

They should recover $slope(t) = 2t$.

Some helpful ideas for computing the slope with a ruler are:

- Use two points that are far from each other rather than two points that are close to each other in order to minimize the error introduced by reading coordinates from the graph.

- Use points which are on the grid if at all possible.

- Coordinates of points on the graph can be found by using the rule for the function.

3. Does the principle "velocity = slope of distance graph" still seem to be true?

Problems

1. Suppose $v(t) = 3 - 4t$ for $0 \leq t \leq 4$. Find the distance function. Graph the distance function. Verify that the slope function for the distance graph is $3 - 4t$.

2. Prove by geometry that the area under the curve $f(t) = 2t$ from $t = 0$ to $t = x$ is given by $A(x) = x^2$.

At this point, based on this activity and "Airplane flight with constant velocity," the class should have following ideas:

- Linear distance corresponds to constant velocity.
- Velocity is the slope of the distance graph.
- Distance is the area under the velocity graph.
- Linear velocity gives a quadratic distance.

Given velocity graph, sketch distance graph

Topics: Graphing

Summary: Students learn a top-down approach to sketching a graph. In particular, they learn how to sketch a distance graph from a velocity graph.

Background assumed: Students need to understand the relationship between distance and velocity; for example, that the (signed) distance increases when velocity is positive, etc.

Time required: 40 minutes

Threads: graphical calculus, distance and velocity, top-down methodology

Comment:
You may want to begin this activity by reminding the students about the top-down approach to problem analysis. See the activity: A formula for a piecewise-linear graph: Top-down analysis.

We will analyze the general problem of sketching the distance graph if we are given the graph of the velocity. We will approach this using the top-down method.

Problem: Given the graph of the velocity, sketch the graph of the distance traveled.

Level I (top level)

When we sketch a graph we draw a curve. There are many different ways to do this. One common method is to draw and scale the coordinate axes, then plot some points, and finally connect these points. However, if the points are arbitrarily selected we may not obtain the correct graph. For instance, if we were drawing the graph of a piecewise linear function and we plotted points that all corresponded to the same linear piece, we might conclude that the graph was a straight line. So, if we want to solve the problem using this method we need to be sure that we plot points that will not lead us into making a mistake.

This approach breaks the original problem into three smaller problems:
I.A. Scaling the coordinate axes
I.B. Choosing and plotting points on the distance graph
I.C. Connecting the points that we plotted

Level II (smaller problems we need to solve in order to complete level I)

First we will analyze A and B. When we choose the scale for the coordinate axes we want to make sure that we "see" all the meaningful information. For instance if we graphed $y = \sin(x/100)$ using $x = 0, 1, 2, 3$, and 4 we would not get the correct idea at all. (Plot the five points and see if you would have guessed the correct shape of the graph.) We also need to know the possible values of the second coordinate in order to decide on an appropriate scale for the vertical axis. So we will have to investigate the maxima and minima of the distance.

We know that when the velocity is positive then the distance traveled is increasing, and when the velocity is negative then the distance traveled is decreasing. So the distance will be a local maximum when the velocity changes from positive to negative. Using the same reasoning, the distance will be a local minimum when the velocity changes from negative to positive. So, we want to plot points on the distance graph that correspond to times when the velocity graph changes sign. There are only two ways for the velocity to change sign:

1. The velocity graph passes through the horizontal axis. When this happens the velocity is equal to 0.

2. The velocity is discontinuous at a time with the velocity positive on one side of the point of discontinuity and negative on the other side of the point of discontinuity.

Times when either of these occurs are called *change points of the velocity*.

The instructor should draw some examples of velocity curves to illustrate both types of change points.

Since we are given the velocity graph we can easily pick out all the times when the velocity changes sign. The points corresponding to these times should be plotted on the distance graph. In addition, since these points will include all the local maxima and minima they will give us the information we need to choose the scale of our graph.

So at level II we need to perform the following tasks:
II.A. Find all the times when the velocity is 0.
II.B. Find all the times when the velocity is discontinuous.
II.C. Classify each interval between succesive change points as an interval where the distance is increasing or as an interval where the distance is decreasing.

Using just this much information we could sketch the rough shape of the distance graph since we know where the distance increases and decreases. However, if we know whether the graph curves up or down (is concave up or concave down) we can make a more accurate sketch. Remember, when the velocity is increasing then the distance graph will be concave up and when the velocity is decreasing then the distance graph will be concave down.

So investigating where the distance graph is concave up or down will be done at level III.

Level III (simpler tasks to complete level II)

It is easy to read where the velocity is increasing and where the velocity is decreasing from the velocity graph. Times when the velocity changes from increasing to decreasing or from decreasing to increasing correspond to points where the distance graph will change from concave up to concave down or vice versa so we will also plot these points on our distance graph. So at level III our tasks are:

III.A. Find all the times when the velocity changes from increasing to decreasing or from decreasing to increasing. These points are called *inflection points*.

III.B. Decide whether the distance graph is concave up or concave down on every interval determined by successive inflection points.

Assembling the solutions

Now we will assemble the results of our simple problems into a solution to the original problem.

First, mark each change point and inflection point on the horizontal (time) axis.

On each interval determined by in this way, the distance graph must be one of the following types of curves

Increasing and concave up

Increasing and concave down

Decreasing and concave up

Decreasing and concave down.

Determine which type the distance graph is on each interval.

Now we know how to sketch the graph between the points chosen in above. So if we plot these points, then we will be able to sketch the distance graph accurately. We have previously discovered that the distance traveled between $t = a$ and $t = b$ is the area under the velocity graph between $t = a$ and $t = b$. Therefore, if we know the distance at any point, we can estimate the distance at any other point by geometry if the velocity graph is straight or by counting blocks on the velocity graph. So, if we know what the distance is at one time, then we can plot all the points we found in the first step on the distance graph.

Next, plot the points on the distance graph corresponding to the times specified in the first step.

Finally, to finish the problem, connect the points plotted in the previous step according to which of the four types of curves describes the distance in that interval.

Problems

1. For each of the following velocities:

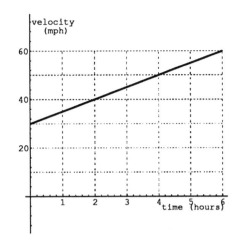

A

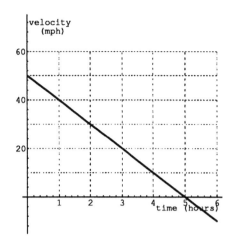

B

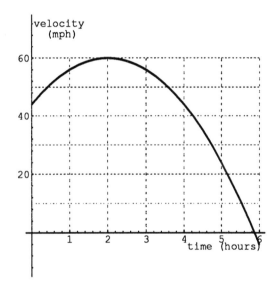

C

(a) Find the distance traveled from $t = 1$ to $t = 3$.

(b) Find the average velocity from $t = 1$ to $t = 3$.

(c) Sketch a constant velocity curve equal to the average velocity from $t = 1$ to $t = 3$.
(d) Find the distance traveled from $t = 1$ to $t = 5$.
(e) Find the average velocity from $t = 1$ to $t = 5$.

(f) Sketch a constant velocity curve equal to the average velocity from $t = 1$ to $t = 5$.
(g) At what time between $t = 0$ and $t = 6$ is the distance from the starting point the greatest?
(h) Find the distance traveled from $t = 1$ to $t = x$.
(i) Sketch the graph of distance traveled versus time.

Note that in order to sketch the distance graph, students need to know the distance at some time. Most students will assume that the distance at time 0 is 0. A good method of illustrating that all the possible distance curves have the same shape is to give different students different initial values, and compare their results.

---

2. Suppose the temperature at sea level is 200° K and the rate of change of temperature with respect to altitude is $100 - 20t°$ K/km at an altitude of t kilometers above sea level.

   (a) Find the change in temperature between sea level and 5 km above sea level.

   (b) Find the temperature at 5 km above sea level.

   (c) At what altitude is it the warmest?

   (d) Sketch the graph of temperature versus altitude.
   (e) Indicate the average temperature between 0 and 5 km.

Homework follow up

1. Go over what happens when velocity changes sign. Be sure they understand that the distance traveled is always the area between the velocity curve and the horizontal axis with the sign of the area positive when velocity above horizontal axis and negative when below horizontal axis.

2. They will need to count blocks to get the area under the velocity graph in part C.

3. After working out the average velocity and average temperature you should be able to state the graphical mean value theorem: If the velocity is continuous between $t = a$ and $t = b$ then the average velocity is equal to the velocity at some point c in (a, b). You can restate this as $f'(c) = (f(b) - f(a))/(b - a)$.

Function-derivative pairs

Topic: Derivative

Summary: Students use the relationship between the graphs of f and f' to identify which graph is which.

Background assumed: Geometric properties of derivatives

Time required: 10 minutes

Thread: graphical calculus

In each figure below, a function and its derivative are shown on the same set of coordinate axes. For each figure, identify which graph is the function and which is the derivative, and explain in one or two sentences what properties of the graphs led you to your choice.

If students have been doing graphical calculus all along, they should not have much trouble with this exercise. It reinforces ideas about increasing and decreasing functions, extrema, concavity, and points of non-differentiability.

More airplane travel

Topic: Distance from velocity

Summary: Students use the relationship between distance and velocity to sketch the distance graph from the velocity graph.

Background assumed: "Airplane flight with constant velocity"

Time required: 40 minutes

Threads: distance and velocity, modeling, graphical calculus

Let's look again at the plane that passed over Ithaca at 1 o'clock and traveled at 500 mph from 1 to 5. At 5 o'clock it enters the air control region for an airport and starts to slow down. It gradually slows down to 300 mph by 5:30.

1. Draw the velocity graph.

Students will usually get this.

---

2. Now draw the distance graph.

Be ready to show why distance still increases.

---

3. Now the plane lands at 5:45. Draw velocity and distance from 1 to 6.

They should have horizontal distance after 5:45.

Two important concepts we will use to analyze graphs are the ideas of *increasing* and *decreasing*. A graph is *increasing* on the interval (a, b) if the graph rises as we move from a to b along the horizontal axis.

4. Write a similar statement describing what it means to say that the graph is decreasing on the interval (a, b).

The instructor should draw some graphs on the board. Include at least one graph that is not monotone and one non-constant function that is constant over some interval. Many students have a problem classifying a horizontal line. One can use the terminology neither increasing nor decreasing. Some instructors may prefer to use the terms non-decreasing and non-increasing.

The distance graph of the plane increased from 1 to 5:45. The velocity graph decreased from 5 to 5:30. Next let's think of a different story. What if the plane developed engine trouble and turned back towards Ithaca at 2:00?

5. Now what would the distance graph look like?

Make sure the class understands the graph of this last example. In particular, everyone should believe that the graph will be decreasing after 2 o'clock since the plane is now getting closer to Ithaca. You may also wish to discuss the shape of the graph at 2: Should it have a corner or should it be "smooth"?

---

6. Now try to draw the velocity graph for the last example.

You should discuss negative and positive velocity here. Make sure everyone agrees that it makes sense to call velocity positive if the plane is heading away from Ithaca and negative if the plane is heading towards Ithaca. Indicate that the distance is increasing when the velocity is positive and the distance is decreasing when the velocity is negative. Point out that plane is furthest from Ithaca when velocity changes from positive to negative. (The use of the rate graph to locate regions where a graph is increasing or decreasing and to locate extrema of the quantity being graphed is a theme that occurs throughout calculus.)

7. Now try to draw the distance graph of the original plane (the plane that did not return to Ithaca) if the plane circled the airport from 5:30 to 5:45.

The instructor should point out two things about this example.

1. The relation of the part of the graph when the plane was circling the airport to graphs of trigonometric functions. Some students may remember that trigonometric functions are called circular functions.

2. This example (the distance graph of the plane) is an example that would be very hard to find a formula to describe. But it occurs in real life, and we can describe it graphically.

Problems

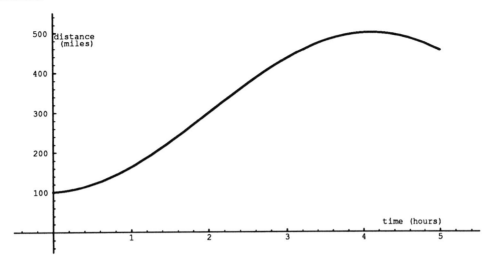

Airplane trip

1. Here is a graph of the distance for another airplane trip. Mark where the velocity is fast and where it is slow. (You can think of the "fast" periods of time as times when the plane enjoys a tailwind, and the "slow" periods as times when there is a headwind.)

2. Dallas to Houston

Dallas to Houston

Topics: Distance and velocity

Summary: Students derive information about the distance function from the velocity function.

Background assumed: Relationship between velocity and distance (i.e., that the distance function is the "area graph" for the velocity graph). The "More airplane travel" activity is useful.

Time required: Homework

Threads: graphical calculus, distance and velocity

The four graphs below each give models for the velocity of a car driving along a straight highway. At time $t = 0$ the car left Dallas and headed for Houston.

1. For each graph, write an explanation of the situation (in complete sentences) that gives the important information contained in the graph.

2. For each graph, decide when velocity is increasing and when velocity is decreasing.

3. Try to give a formula for the velocity in terms of time.

4. Decide whether you think the car could behave like the model. Explain your reasons.

5. Sketch the graph of distance the car has traveled versus time.

6. Try to find a formula for distance in terms of time.

7. For each graph, decide when the car is furthest away from Dallas. Explain.

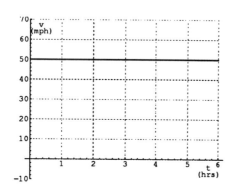

A

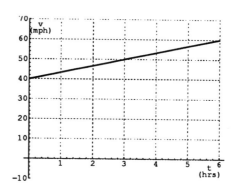

B

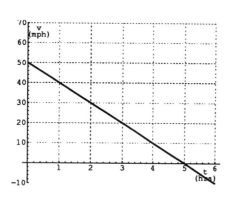

C

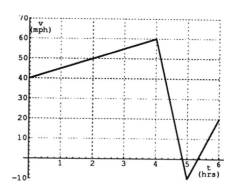

D

Homework followup:

1. Relate the distance traveled to the area under the velocity curve.

2. You can distinguish concave up and concave down by contrasting the second and third distance graphs. At this stage we have also used the terms "curves up" and "curves down," or "bends up" and "bends down."

Water tank problem[2]

Topic: Maximum-minimum

Summary: Students find a maximum for a function given graphically. This can be used to motivate the first-derivative test.

Background assumed: Rates and tangent lines

Time required: 40 minutes

Resources needed: A ruler is helpful.

Threads: graphical calculus, modeling

At time $t = 0$, water begins to flow from a hose into an empty tank at the rate of 40 liters/min. This flow rate is held constant for two minutes, at which time the tank contains 80 liters. At that time, ($t = 2$), the water pressure is gradually reduced, until at time $t = 4$, the flow rate is 5 liters/min. This flow rate is held constant for the final two-minute period, at which time ($t = 6$) the tank contains 120 liters.

1. Draw a graph of the volume V of water in the tank (liters) against the time t (min) for $0 \leq t \leq 6$.

The fact that the pressure is gradually reduced between time $t = 2$ and $t = 4$ means that the graph of V should be concave down in the interval $2 \leq t \leq 4$. There will be several different correct graphs.

---

[2]Adapted from Peter Taylor, *Calculus: The Analysis of Functions*, Wall & Emerson, Inc., 1992.

2. What is the average rate of flow into the tank over the entire six-minute period? Show how this can be interpreted on your graph.

The average rate of flow over the entire six-minute period is the slope of the straight line connecting the initial point and final point on the graph of V.

---

3. Now, suppose that, in addition to the above sequence of events, a pump is started up at time $t = 2$, and, for the next four minutes, pumps water out of the tank at the constant rate of 15 liters/min. What we have called V above now represents the total amount of water that has flowed into the tank at time t, and we let W represent the total amount of water pumped out of the tank at time t. Plot W on the same set of axes. Show how to interpret the volume of water in the tank at any time.

4. Show how to find on your graph the point at which the water level in the tank is a maximum.

The water level rises if the rate at which water enters the tank is greater than the rate at which the water is pumped out of the tank. So, any time the slope of the graph of V is larger than 15 is a time when the water level is rising (increasing). In the same way, the water level is falling (decreasing) if the slope of the graph of V is less than 15. For example, the water level is increasing for $0 \leq t \leq 2$ and decreasing for $4 \leq t \leq 6$. Some time between 2 and 4 the water level will change from increasing to decreasing; at this time the water level will be a maximum. The slope of the graph of V at this time should be 15. An elegant way to find this time is to line up your ruler with the graph of W (which has slope 15) and move the ruler up (keeping the ruler at the same angle) until it is tangent to the graph of V. The time corresponding to this point on the graph of V is the time when the water level is maximized.

---

5. What is the instantaneous flow rate from the hose into the tank at this time?

6. Now on a new set of axes draw the graph of the rate at which water enters the tank (liters per minute) against the time (min) for $0 \leq t \leq 6$. (This is the graph of the rate of change of V with respect to time.)

7. On the same set of axes that you used in 6 plot the graph of the rate at which water is pumped from the tank. (This is the graph of the rate of change of W.)

8. Find any times where the rate graphs that you drew on parts 6 and 7 intersect. What can you say about the level of water in the tank at these times?

Point out that as long as the rate of change for V is above the rate of change for W the water level is rising. The water level is falling when the rate of change for V is below the rate of change for W. So, when the rate of change for V crosses the rate of change for W (from above to below) the water level is a maximum. Using the rate graphs makes it easier to identify the point where the maximum occurs. This is a good time to mention that one of the basic ideas of calculus is the use of rates to investigate quantities.

---

9. If we continue the action past time $t = 6$, with the constant flow in of five liters per minute and the constant flow out of 15 liters per minute, at what time will the tank be empty? Show how to interpret this graphically using the above plot of V and W.

Tax rates and concavity

Topics: Concavity, increasing and decreasing functions

Summary: The idea of a progressive tax rate (an increasing derivative) is used to motivate concavity (of the tax function).

Background assumed: Derivative as a rate of change

Time required: 25 minutes

Thread: graphical calculus

In this activity, we will use some simple ideas about taxes and tax rates to investigate some basic properties about functions.

Economists use the term *marginal tax rate* to refer to the additional tax paid on the next (taxable) dollar earned. For example, suppose that under some tax plan the tax on $34,567 is $6897.76 while the tax on $34,568 is $6898.04. Then, for a taxpayer earning $34,567, the tax paid on the next dollar is $0.28, so the marginal tax rate is $0.28/$1 = 28%. (Note that this is not directly related to $6897.76/$34,567 = 19.95%, the percentage of taxable income that goes to tax.)

Let $t(x)$ denote the amount of tax that is paid on x of taxable income.

1. Economists refer to a tax as *progressive* if the marginal tax rate increases (as taxable income increases). If $t(x)$ is a progressive tax, what will the graph of $t(x)$ look like? Sketch $t(x)$ for a progressive tax.

   The graph of $t(x)$ should be concave up. You can mention that we are using a continuous function to model a discrete situation.

2. At a few points on your graph, sketch the tangent line. What is the relationship between the graph and its tangent lines?

   The graph is above the tangent line.

3. Explain how the marginal tax rate is related to $t'(x)$. (Hint: Take $\Delta x = \$1$ in the definition of $t'(x)$.)

   The marginal tax rate is a close approximation to $t'(x)$. (It's the difference quotient with $\Delta x = \$1$, which is "small" relative to x.)

   Some students will not see this. Remind them of the definition of $t'(x)$.

4. If t has a second derivative, what property will t'' have?

The second derivative, t'', will be positive. (Students need to associate t' with the marginal tax rate; and they need to know that if a function (t') is increasing, then its derivative (t'') is positive.)

A function with an increasing derivative is said to be *concave up*. As you have seen, geometrically this means that the graph lies above its tangent lines; in terms of derivatives, it means that its second derivative is positive.

The definitions above can be "turned upside down"; that is, a function with a decreasing derivative is said to be *concave down*. Geometrically this means that the graph lies below its tangent lines; in terms of derivatives, it means that its second derivative is negative.

Problems

Consider the function $u(x) = x - t(x)$.

1. Describe u in words. That is, what does $u(x)$ represent?

2. What properties does u have? (What assumptions are you making about t?)

3. Is it possible for a function to be positive, concave down, and unbounded (approach infinity)? Explain.

The function $u(x)$ represents what you get to keep if you earn x dollars. Presumably u is positive and increases; it is also concave down.

---

4. Suppose that over some income range (e.g., for all taxable income between $21,500 and $52,000) the marginal tax rate, $t'(x)$, is constant (e.g., 28%). What will the graph of $t(x)$ look like over that range?

If the marginal tax rate is constant over some income range, then t will be linear over that range.

Testing braking performance

Topics: Rate graphs and areas

Summary: Students use velocity to calculate distance traveled in two situations: one where velocity is given by a graph and another where velocity is given by a formula.

Background assumed: Relationship between velocity and distance

Time required: 25 minutes

Threads: graphical calculus, multiple representation of functions. distance and velocity

---

Buyers' Union is testing the braking performance of two cars: the Minima and the Corollari. The test consists of using the brakes to bring the car to a complete stop when the car is traveling at 60 miles per hour. The Minima slows down at the uniform rate of 8 miles per hour per second. The Corollari's velocity is given in the graph below.

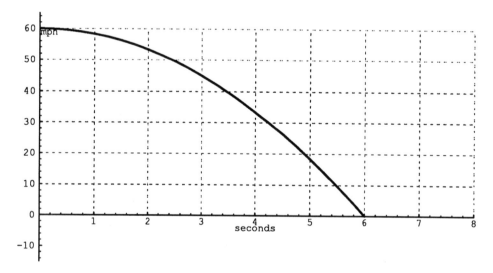

Velocity curve

1. How much time will it take for each car to stop?

2. How far will each car travel before it stops?

3. Write a brief paragraph summarizing braking performance which will be inserted into the report comparing these cars. If your data indicates that one of the cars is superior explain your conclusion.

---

1. Note that the car with the shorter stopping time travels a greater distance.

2. For the Minima, draw a graph of the Minima's velocity on the given graph—that is, a straight line through $(0, 60)$ and $(7.5, 0)$—and compare the areas under the two velocity graphs.

The start-up firm

Topic: Integration

Summary: Students encounter graphical integration (the integral as area) in an economic context.

Background assumed: Relationship between aggregate and rate (area graph from rate graph)

Time required: 35 minutes

Thread: graphical calculus

Pat and Chris are involved in a new business. One of Pat's jobs is to estimate the company's prospects for raising money (finding investors at the beginning, selling their product eventually); and Chris is responsible for spending it (buying equipment, finding office space, hiring staff, ...). Both have been making predictions about how much they will be raising/spending. What is shown on the graph below is a graph of these projections. The horizontal axis represents months; the vertical axis measures dollars per month. (Note that what is illustrated is the *rate* at which money is raised/spent—not the total amount raised/spent.) Both are particularly interested in the company's net worth (the difference between the amount raised and the amount spent).

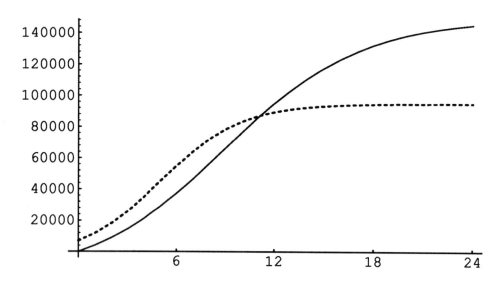

Pat: solid line; Chris: dotted line

It may help the students out if you begin by telling them to mark clearly on the graph which line gives them information about money coming *in* and which one represents money going *out*.

---

1. Estimate the company's net worth in two months.
2. Is net worth positive or negative after six months? after 12 months? after 24 months?

Many students will say that net worth is positive after 12 months. Suggest that they think about the next question.

---

3. When is net worth decreasing? When is it increasing?

Net worth decreases [resp., increases] when the rate in (the solid line) is less than [greater than] the rate out (dotted line). It may help to elaborate: net worth = revenue − cost, so net worth′ = revenue′ − cost′. Consequently, net worth′ > 0 if and only if revenue′ > cost′. (Intuitively this comes as no surprise to students; the "formal" explanation confirms their intuition.) Here, net worth decreases for (about) the first 12 months and increases after.

4. When is net worth smallest? largest? zero?

From considering 3, net worth is smallest at (about) 12 months (when the graphs cross). Emphasize that net worth is smallest when it stops decreasing and starts increasing. Net worth ultimately increases (at almost a constant rate), so there is no maximum. Net worth is zero at about 17 months (when the areas are equal).

5. Sketch a graph of net worth.
6. Both graphs seem to flatten out. Interpret this phenomenon. (What does this tell you about net worth?)

Net worth will increase at (almost) a constant rate.

Graphical composition

Topic: Composition of functions

Summary: This activity can be done while introducing (or reviewing) composition. One point of the activity is to demonstrate a technique for finding values for the composition of two functions when the component functions are represented by graphs.

Time required: 20 minutes.

Threads: graphical calculus, modeling

Below is a sketch of the graphs of two functions $y = f(x)$ and $y = g(x)$. Given a point x_0 in the domain of f with the property that $g(x_0)$ is in the domain of f, there is a geometric method that can be used to plot $(x_0, f \circ g(x_0))$. The method determines a polygonal path (a path made up of straight line segments) that starts at the point $(x_0, 0)$ on the x-axis and ends at $(x_0, f \circ g(x_0))$. We illustrate the construction of this path below.

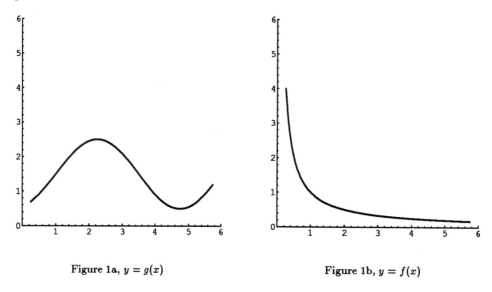

Figure 1a, $y = g(x)$ Figure 1b, $y = f(x)$

First, superimpose the graph of f on the graph of g and also sketch the graph of the line $y = x$.

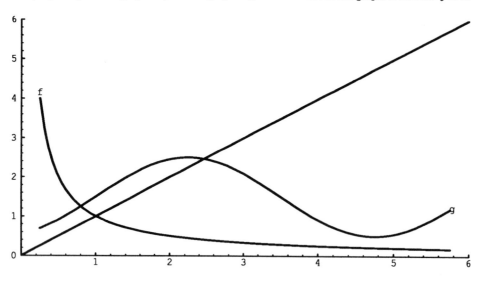

Figure 2

Next, starting at the point $(x_0, 0)$, travel along the line $x = x_0$ until you meet the graph of g at $(x_0, g(x_0))$. This determines the first leg of the path. (See Figure 3.)

Note: The illustrations assume that $x_0 = 3$.

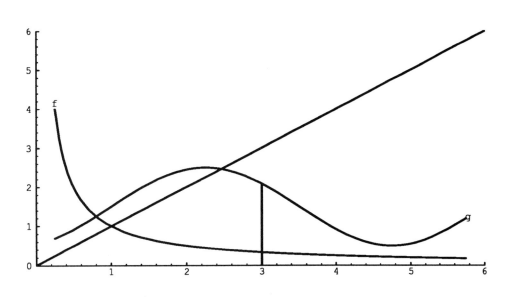

Figure 3

Now, travel along the line $y = g(x_0)$ until you meet the line $y = x$ at the point where the x and y coordinates are equal. This is the point $(g(x_0), g(x_0))$. This determines the next leg of the path. (See Figure 4.)

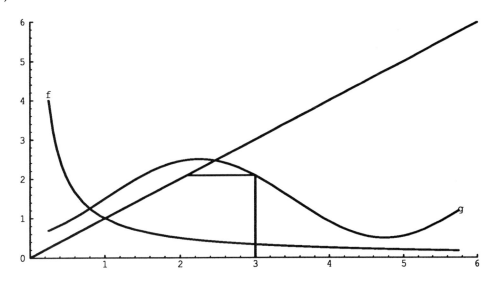

Figure 4

Next, travel along the line $x = g(x_0)$ until you meet the graph of f at $(g(x_0), f(g(x_0)))$. This determines the third leg of the path. (See Figure 5).

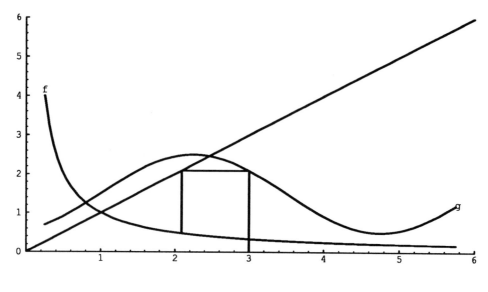

Figure 5

Finally, travel along the line $y = f(g(x_0))$ until you meet the line $x = x_0$ at $(x_0, f(g(x_0)))$. This is the last leg of the path and it terminates at $(x_0, f(g(x_0)))$, the desired point. (See Figure 6.)

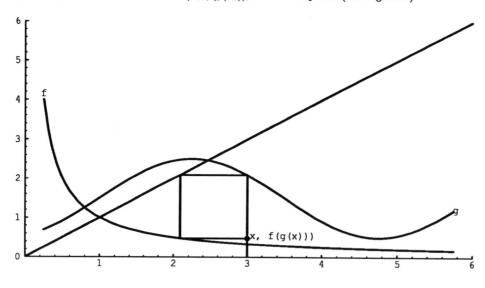

Figure 6

To recap, start at $(x_0, 0)$, travel along $x = x_0$ to the graph of g, travel to the line $y = x$, travel to the graph of f, travel back to the line $x = x_0$.

For the functions graphed above:

1. Find $(f \circ g)(2)$.

2. Find $(f \circ g)(4)$.

Problems

In this example, the graph of f represents the number of ten-gallon containers of fresh milk a certain farmer can expect to sell per day on the open market as a function of the total amount of milk available in a particular metropolitan market. The total milk is measured in thousands of gallons. For example, if only 250 gallons are available (0.25 thousand gallons), the farmer could expect to sell 4 ten-gallon containers; i.e., 40 gallons.

The graph of g represents the number of gallons (measured in thousands) as a function of time over the six months where one represents the end of January, two represents the end of February, six represents the end of June, etc.

1. (To be done individually.) Estimate how many gallons of milk the farmer was able to sell on the last day of January, of February, of March, and of June.

2. (Group) Can you give a sketch of the farmer's selling potential over the time scale $[1,6]$? Give a description of what this graph says about the farmer's selling potential.

This activity contains a basis graphical method for computing compositions that is an extension of the "cobweb" diagram for iterations. The instructor could use the narrative as an introduction, have students practice 1 individually, and then have groups discuss the second question.

The context of the problem should help in question 2. As the total amount of milk goes up, the farmer's potential goes down and vice versa. This is logical. There are general results that the students can abstract about $f \circ g$, if f is a positive-valued, decreasing function, for example. There should be some discussion about the fact that the domains of f and g don't match. (It's more important that the range of g be contained in the domain of f.)

"Cobweb" diagrams occur when $f = g$. That could form an extension of this activity. That is, ask what is the diagram for $f \circ f(x_0)$? for $f \circ f \circ f(x_0)$?

After the last problem has been completed, the instructor can ask the students if they notice any general pattern regarding the shape of the composition. Note here that the function f is decreasing, so they may notice that $f \circ g$ is increasing [decreasing] when g is decreasing [increasing]. The scenario in the problem is meant to support this observation: the farmer's selling potential decreases as the number of gallons on the market increases, etc. (So, here, the context helps students understand the theory.)

The leaky balloon

Topic: Related rates

CHAPTER 1. GRAPHICAL CALCULUS

Summary: This example provides a graphical approach to a related rate problem.

Background assumed: Derivatives

Time required: 15 minutes

Threads: graphical calculus, multiple representation of functions

---

A spherical balloon is leaking. Below is a graph of the radius (r) as a function of time (t).

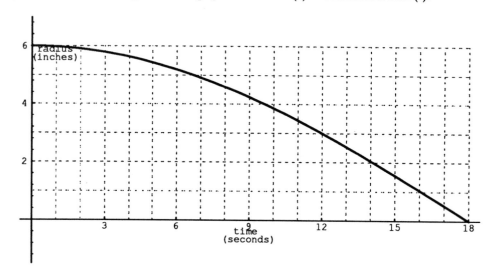

Leaky balloon

1. How fast is the radius changing when $t = 15$?

2. How fast is the radius changing when the radius is 4 inches?

3. How fast is air escaping (i.e., how fast in the volume changing) when the radius is 4 inches? [$V = \frac{4\pi r^3}{3}$.]

Problem
   How fast is air escaping when $t = 15$?

Inverse function from graphs

Topic: Inverse functions

Summary: This activity can be used to introduce the geometry of inverse functions.

Background assumed: It is helpful to have done the activity "Graphical composition" first.

Time required: 20 minutes

Thread: graphical calculus

Figure 1 refers to a graph of the velocity of an airplane during the first phase of its trip. Here, the plane is traveling at different velocities at different instants. So, if we had a record of the velocities, we could theoretically pinpoint the one and only one instant that the plane was traveling at that velocity. The function that determines the time from the velocity is called the *inverse* of the function v and is denoted v^{-1}. So, in the terminology of inverse functions, for each time t_0, $v^{-1}(v(t_0)) = t_0$, and for each velocity v_0, $v(v^{-1}(v_0)) = v_0$. Obviously v would not have an inverse if the plane took on the same velocity at more than one instant, for in that case we couldn't determine a unique time from that velocity.

Using the fact that $v^{-1}(v(t_0)) = t_0$ and our geometric results on composition, we can sketch the graph of v^{-1} on the *same* axis as the graph of v.

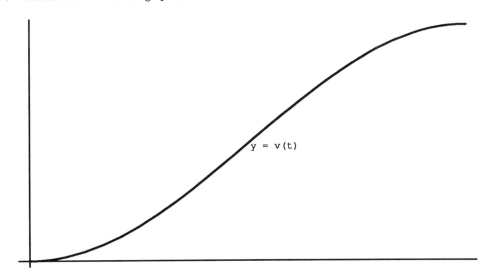

Figure 1

Note that for the function v, the x–axis is interpreted as a time axis, while when we consider the function v^{-1}, the points on the x–axis measure velocity.

1. Give an argument that supports the claim that given v_0, the end of the polygonal path in Figure 2 will be the point $(v_0, v^{-1}(v_0))$.

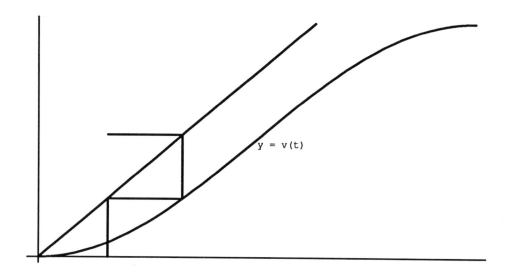

Figure 2

2. Give an argument that supports the claim that the box in Figure 3 is a square.

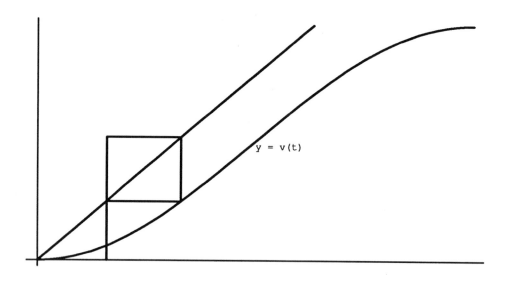

Figure 3

3. Give an argument that suggests that (v_0, t_0) is the reflection of the point (t_0, v_0) about the line $y = x$.

4. Using the results of 3, sketch a graph of v^{-1} on the same axes in Figure 1. (Plot a few points of v^{-1} first.)

Problems

1. Sketch the graph of an increasing function $y = f(x)$.

2. Sketch a graph of f^{-1}.

3. Sketch a graph of $f^{-1} \circ f$.

Chapter 2

Functions, Limits, and Continuity

Introduction to functions

Topics: Function, domain, range

Summary: Introduces the mathematical concept of function. This activity can be used by students while the included materials are being presented or discussed in class.

Time required: 30 minutes.

Thread: multiple representation of functions

One of the most useful ideas in mathematics is the concept of function. So far in these pages many examples of functions have occurred, though we have not especially stressed the properties that distinguish the examples as functions.

A *function* is a rule that associates with each element of some set, called the *domain* of the function, exactly one element of another set, called the *co-domain* of the function. If the name of the function is f, if its domain is X, and if its co-domain is Y, then we convey these facts by using the notation

$$f: X \to Y.$$

In first-year calculus most functions have as both domain and co-domain the set **R** of all real numbers, or some subsets of **R**.

The way in which the function associates elements of the domain with elements of the co-domain can be conveyed in a variety of ways: by a graph, in words, by a table or a formula; functions often arise from physical phenomena. The important thing to remember is that in order to be a function, for *each element* of the domain X, there must be *exactly one element* of the co-domain Y.

For example, consider the graph of Josh's interrupted trip to the library, in Library Trip. It represents a function, which we will call G. The graph is reproduced here.

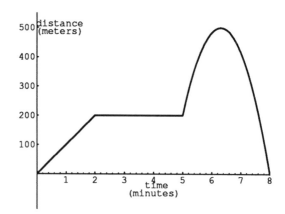

Library trip

When a function is presented on a graph, it is customary to represent the domain of the function on the horizontal axis and to represent the co-domain on the vertical axis. The domain is the part of the horizontal axis that corresponds to points actually included in the graph. In this example, the domain of G consists of all the numbers between 0 and 8. In set notation, we would write

$$\text{domain of } G = \{x \in \mathbf{R} \mid 0 \leq x \leq 8\}$$

We can think of the co-domain as the set $\mathbf{R}$ of all real numbers. However, not every real number on the vertical axis corresponds to a point on the graph. The elements of the co- domain that actually correspond to points on the graph make up the *range* of the function. In this case, the range of the function G is the set

$$\text{range of } G = \{y \in \mathbf{R} \mid 0 \leq x \leq 500\}$$

To determine what element of the range is associated with a specific element of the domain, we would read the value from the graph. Since the points $(2, 200)$ and $(6, 483)$ lie on the graph, we see that the function G associates with the number 2 (in the domain) the number 200 (in the range), and with the number 6 (in the domain) G associates the number 483 (in the range). It is customary to use more concise notation to express these facts. For the two facts above, we would write

$$G(2) = 200 \quad \text{and} \quad G(6) = 483$$

1. Use the graph to find (or estimate) $G(4)$, $G(0.7)$, $G(5.9)$, and $G(0)$.

2. Given that a function must associate exactly one range element with each domain element, how can you tell by looking at a graph whether or not the graph represents a function?

Many students will already be familiar with the vertical line test. The others can "discover" it here.

---

3. Which of the graphs below represent functions? Explain.

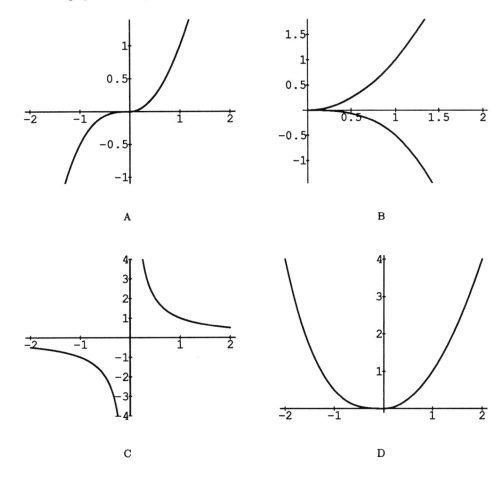

A

B

C

D

4. If T is a function that is presented as a table, we read the function values directly from the table. For example, the table at the right represents a function. The domain is the set of integers $\{1, 2, 3, 5\}$.

x	$T(x)$
1	13
2	4
3	-6
5	0.1

For this function, you can read the values right off the table. For example, $T(3) = -6$. Find $T(1)$, $T(2)$, and $T(5)$.

Notice that it doesn't make sense to ask for $T(4)$ or $T(8.7)$ because 4 and 8.7 are not in the domain of the function. Also notice that if we added a line "2   7" to the table we would have violated the "one and only one value" rule by trying to have two different values associated with 2.

5. Some functions are presented as formulas. If this suits the situation, this can be very useful, because there are lots of well understood methods of analyzing the properties of such functions. For example, the formula $f(x) = 3x + 4$ defines a function. Unless explicitly stipulated, the domain consists of all the values of x that make sense in the formula. The domain in this case is all real numbers. For any domain element, the formula tells us the way to calculate a range element. Thus, $f(-0.8) = 1.6$ (i.e., $3 \times -0.8 + 4$). Find $f(2)$, $f(0.06)$, $f(-\pi)$.

Mention that the dependent variable is a "dummy variable." Ask the students to find $f(z)$, $f(u^2)$, and $f(x+1)$.

For such functions, we usually write $f(x)$ = formula in x. This notation tells you the name of the function (in this case, f) and helps clarify what the domain is. (As mentioned above, unless otherwise specified, the domain consists of all the values of x that make sense in the formula on the right side of the equals sign.) We obtain specific function values, as we did above, by substituting for x on both the left and the right side.

One way of visualizing a function is as a machine. The machine accepts inputs (values from the domain), does whatever it was designed to do, and then delivers outputs (members of the co-domain).

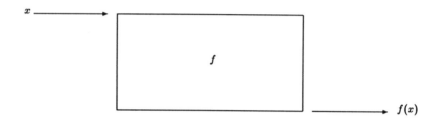

Inside the machine can be a formula, a table, a graph, or some other mechanism. As long as the machine produces exactly one output for every (valid) input, it represents a function.

Problem
Find the domain and range for all the functions you found in 3.

Postage

Topics: Continuity

Background assumed: Piecewise-defined functions

Summary: Postage is used to illustrate the concept of continuity. This leads naturally to the concept of the one-sided limit.

Time required: 10–15 minutes

Thread: graphical calculus

You are in charge of a mail order company. Your company sends out 10,000 letters each day. The post office rates for postage are:

29 cents for a letter that weighs 1 ounce or less,

54 cents for a letter that weighs more than 1 ounce but not more than 2 ounces,

77 cents for a letter that weighs more than 2 ounces but not more than 3 ounces, and

98 cents for a letter that weighs more than 3 ounces but not more than 4 ounces.

1. How much will it cost to mail a letter that weighs 1 ounce? 1.5 ounces? 0.8 ounces?

A few students will make the postage proportional to the weight, although the definition specifically says this is not the case. Be sure the students know that the postage for 1 ounce is the same as the postage for 0.8 ounces.

---

2. Sketch a graph of the postage for a letter as a function of the weight of the letter for weights from 0 to 4 ounces.

Some students will want to connect the discontinuous parts of the graph. You should point out that the resulting graph is not the graph of a function.

The postage function is discontinuous at each integer in its domain. The rest of the activity illustrates the importance of discontinuities. You can introduce or review left- and right-hand limits as well as continuity using this activity. The limits can be computed from the graph or by using the values of the function.

---

3. The catalog department plans to send out 10,000 letters that each weigh .85 ounces. How much will the postage cost?

4. The marketing department wants to add an insert to these letters. The insert weighs 0.2 ounces. If you agree to add the inserts to the letters, what will happen to the cost of postage for the letters?

5. The book department is sending out 5,000 letters that each weigh 0.75 ounces. What will happen to the price of the book department's mailing if you agree to add the marketing department's insert to each letter?

6. Summarize what the graph in 2 means to the head of the mail order company.

Problems

1. A long distance telephone call costs $1.25 for the first minute or less and $0.50 for every additional 30 seconds or part of 30 seconds after the first minute.

    (a) How much will it cost to talk for 2 minutes? for 2 minutes and 5 seconds? for 2 minutes and 25 seconds?

    (b) Draw the graph of the cost of a long distance call as a function of the length of the call for calls that last from one second to four minutes.

    (c) Point out any discontinuities on your graph and explain what they mean to a student who calls long distance every night.

2. A parking lot charges $2 for the first hour or less and then $1.50 for any hour or fraction of an hour after the first hour.

    (a) Sketch the graph of the parking charge as a function of how long a car remains in the lot for any time up to 5 hours.

    (b) Point out any discontinuities on your graph and explain what they mean to someone who parks in the lot every day.

What's continuity?

Topic: Continuity

Summary: Students develop the definition of continuity at a point by considering various points on a graph: one point of continuity and points where each of the three conditions (existence of $f(c)$, $\lim_{x \to c} f(x)$ exists, and equality) fail.

Background assumed: Limits

Time required: 10 minutes

Thread: graphical calculus

If you look at the graph of the function below, you can probably say exactly where the function f is and is not continuous.

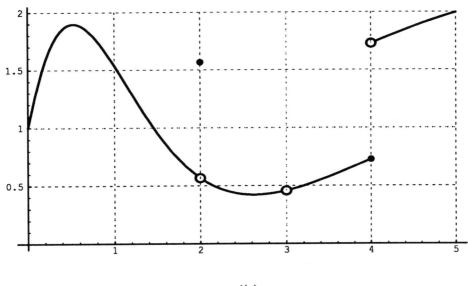

$$y = f(x)$$

1. For what values of x is f *not* continuous?

   _____

   Be sure the class is in agreement before proceeding.

   _____

2. For each of the following values of c, indicate $\lim_{x \to c} f(x)$, $f(c)$, and whether f is continuous at $x = c$.

   (a) $c = 1$

   $\lim_{x \to 1} f(x) =$ _____

   $f(1) =$ _____

   Is f continuous at $x = 1$? _____

   (b) $c = 2$

   $\lim_{x \to 2} f(x) =$ _____

   $f(2) =$ _____

   Is f continuous at $x = 2$? _____

   _____

   Some students may get the limit wrong here.

   _____

(c) $c = 3$

$\lim_{x \to 3} f(x) =$ _____

$f(3) =$ _____

Is f continuous at $x = 3$? _____

(d) $c = 4$

$\lim_{x \to 4} f(x) =$ _____

$f(4) =$ _____

Is f continuous at $x = 4$? _____

3. Using the language of limits, what does it mean for a function f to be continuous at c?

The class will probably identify all three conditions: the limit exists, the function value exists, these two are equal.

Limits and continuity from a graph

Topics: Limits, continuity, and differentiability

Background assumed: Students should have had an intuitive introduction to one- and two-sided limits, continuity, and derivatives.

Time required: 15 minutes

Thread: graphical calculus

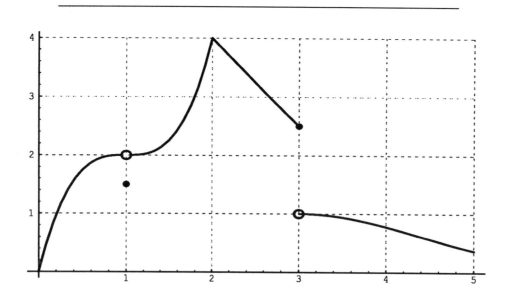

Graph of the function $y = f(x)$

1. Find the value of each of the following. If the value does not exist (if it is undefined), write **DNE**.

    (a) $\lim_{x \to 1^+} f(x)$

    (b) $\lim_{x \to 1^-} f(x)$

    (c) $\lim_{x \to 1} f(x)$

    (d) $f(1)$

    (e) $\lim_{x \to 2^+} f(x)$

    (f) $\lim_{x \to 2^-} f(x)$

    (g) $\lim_{x \to 2} f(x)$

    (h) $f(2)$

    (i) $\lim_{x \to 3^+} f(x)$

    (j) $\lim_{x \to 3^-} f(x)$

    (k) $\lim_{x \to 3} f(x)$

    (l) $f(3)$

    (m) $\lim_{x \to 4^+} f(x)$

    (n) $\lim_{x \to 4^-} f(x)$

    (o) $\lim_{x \to 4} f(x)$

    (p) $f(4)$

Some students may have difficulty at 1 and 3.

---

2. True or false, with reasons:

    (a) The function f is continuous at $x = 1$.

    (b) The function f has a derivative at $x = 1$.

    (c) The function f is continuous at $x = 2$.

    (d) The function f has a derivative at $x = 2$.

    (e) The function f is continuous at $x = 3$.

    (f) The function f has a derivative at $x = 3$.

CHAPTER 2. FUNCTIONS, LIMITS, AND CONTINUITY

(g) The function f is continuous at $x = 4$.

(h) The function f has a derivative at $x = 4$.

The whole point of this activity is for the student to learn what continuity and differentiability "look like." Point out that a function may be continuous but not differentiable (as this one is at 2).

Slopes and difference quotients

Topics: Difference quotients

Summary: Students interpret difference quotients on a graph.

Background assumed: Secant lines and tangent lines

Time required: 15–20 minutes

Resources needed: A ruler is helpful.

Threads: graphical calculus

Answer the following questions about the function f, whose graph is given below.

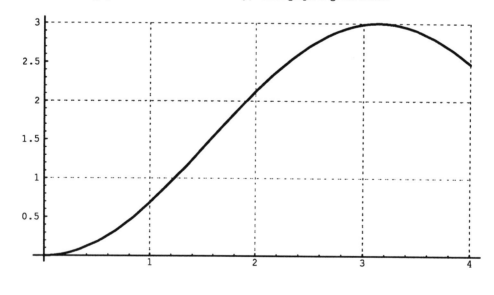

$y = f(x)$

1. Find $f(3)$.

2. Find $f'(3)$.

3. What is the average rate of change of f over the interval $[0, 3]$.

Sketch on the graph a line whose slope equals:

4. $\dfrac{f(3) - f(2.5)}{3 - 2.5}$

5. $\dfrac{f(3)}{3}$

6. $\lim\limits_{h \to 0} \dfrac{f(3+h) - f(3)}{h}$

7. $\lim\limits_{h \to 0} \dfrac{f(h)}{h}$

Mark a point on the curve where the x-coordinate has the property:

8. $\dfrac{f(x)}{x} = 1$

9. $\lim\limits_{h \to 0} \dfrac{f(x+h) - f(x)}{h} = 1$

This activity reinforces the geometric interpretation of the concepts of function, limit, derivative, and difference quotient. In particular, it allows the students to see graphically a number of interpretations of the concept of average rate of change of a function. An interesting question to ask is, "What is the largest value that $f(x)/x$ can assume on the interval $[0, 4]$?"

Sequences

Topics: Sequences and iteration

Summary: Students construct sequences and investigate convergence using a calculator to iterate a function.

Background assumed: Depending on when sequences are introduced in the course, this activity could be used to introduce the very idea of sequences, convergence, and divergence. However, it is probably most effective when question 4 can be done, in which case the derivative as the limit of difference quotients is needed along with the derivatives of x^2 and $\sqrt{x}$.

Time required: 15 minutes

Resources needed: Calculator

Thread: approximation and estimation

---

1. Using the $\boxed{x^2}$ button on your calculators, the group should construct terms of several sequences using the following procedure. Each member of the group should chose a different number and record it. This number should be considered the first term of a sequence and entered into the calculator. The $\boxed{x^2}$ button should be pushed several times and each result recorded as the next term of the sequence until a pattern of convergence or divergence develops. If the sequence appears to be approaching a limit, specify the limit. If the sequence appears to be diverging indicate this also. Each member of the group should repeat this process with two *additional* initial numbers. Now, as a group, consider the following questions. When will you get a convergent sequence? How many different limits are there? What happens to the difference between successive terms when the sequence converges? diverges?

2. Repeat the above procedure using the $\boxed{\sqrt{x}}$ button on the calculator.

3. Discuss the similarities and differences in results obtained in 1 and 2 above.

4. You have constructed several terms of several sequences associated with functions $f(x) = x^2$ and $g(x) = \sqrt{x}$. For the sequences that diverged, can you find a relationship between the derivative and the difference between successive terms of the sequence? What about a relationship between the derivative of the function at points of a convergent sequence and the difference between successive terms of the sequence? Is there any relationship between the derivative of a limit point of a sequence and the kind of sequences that converge to it?

This activity can be done to experiment with convergence using an inexpensive calculator. It has been used when first introducing sequences. It can also be used to introduce recursion. There is a direct connection between convergence and differential calculus. (So, sequences do relate to things the students have studied up to this point.)

If a sequence that has been constructed using the algorithm in this activity converges to a point x_0, then $|f'(x_0)| \leq 1$. With some handwaving, a look at points in the sequence close to the limit point, say x_n and x_{n+1}, and their iterates, $f(x_n)$ and $f(x_{n+1})$, suggests that the size of the difference quotient $\frac{|f(x_n)-f(x_{n+1})|}{|x_n-x_{n+1}|}$ should be less than or equal to 1, also. This in turn means that $|f(x_n)-f(x_{n+1})| \leq |x_n-x_{n+1}|$. That is, subsequent points in the iteration don't spread out. (This is a nod to the fact that a convergent sequence is a Cauchy sequence, but it is only a nod!) Although this is not a rigorous proof, it does emphasize one interpretation (and importance) of the derivative. Moreover, the size of the derivative at points of a divergent sequence is greater than one in these cases. A similar analysis of the difference quotient at these points tends to suggest that subsequent points in this sequence get further apart. It may be noted that sequences that are constructed using this iteration method converge to a fixed point of the function. Moreover, the topics of attracting fixed points and repelling fixed points could be discussed, also. Note that all sequences other than the sequence that starts with zero will converge to 1 under the function $f(x) = \sqrt{x}$.

Can we fool Newton?

Topic: Newton's method revisited

Summary: This activity can be done after Newton's method has been discussed briefly and the students understand the geometry (below) of how the sequence $\{x_n\}$ is determined. The intent is to explore some of the phenomena that may occur and trigger the students' imagination and creativity.

Background assumed: Newton's method

Time required: 20 minutes

Resources needed: Calculator (or graphing software)

Thread: approximation and estimation

In class you have discussed Newton's method for finding a zero of a given differentiable function f. The method consists of computing terms of a sequence $\{x_n\}$ by choosing an initial value x_0 and determining the values x_1, x_2, x_3, and so on using Newton's "formula." If the sequence $\{x_n\}$ converges to a point x_* in the domain of the function, the point x_* might be a root of the equation $y = f(x)$. This activity explores some of the questions concerning the sequence $\{x_n\}$. Most questions deal with cases when things go wrong.

1. If one exists, sketch a portion of the graph of a differentiable function f such that there is a point x_0 where the sequence $\{x_n\}$ determined using Newton's method diverges to infinity. Give a broad description of your function. If no such function exists, support this conclusion.

2. If one exists, sketch a portion of the graph of a function f for which there is exactly *one* point x_0 in the domain of f such that the corresponding sequence $\{x_n\}$ determined by Newton's method converges. If no such function exists support your conclusion.

3. Suppose $\{x_n\}$ has been determined by Newton's method.

    (a) If $x_3 = x_0$, discuss whether or not $x_4 = x_1$.

    (b) Sketch a portion of the graph of a function for each of the following:

    i. There is a point x_0 with $x_0 = x_2$, but $x_0 \neq x_1$.
    ii. There is a point x_0 with $x_0 = x_3$, but $x_0 \neq x_2$.

    (c) Describe a general process for determining a function f that has the property that there is a point $x_0 = x_n$, but $x_0 \neq x_{n-1}$, for $n = 4, 5, 6, \ldots$

We recommend that you discuss parts 1–3 before making this assignment. It is hoped that students have already experienced the fact that polynomials have a finite number of "bumps" and trigonometric functions have "discrete bumps," as well as how exponential functions ($y = c\,e^x$) look. Note that 1 will probably raise questions like: if $x_0 = x_3$ and $x_0 \neq x_2$ then can $x_0 = x_1$? The students ought to be able to answer this at this point.

Problem

Having had the experiences 1–3, discuss the types of behavior that might happen to a sequence $\{x_n\}$ determined by Newton's method for a differentiable function f. What behavior would you expect if x_0 is randomly chosen and f is a trigonometric function? if f is an exponential function?

Chapter 3

Derivatives

Linear approximation

Topic: Derivative as linear approximation

Summary: Students zoom out to see that the tangent line is the line that best fits the graph of a function.

Background assumed: "Graphical estimation of slopes"

Time required: 15 minutes

Resources needed: Graphing calculator or software

Threads: graphical calculus, approximation and estimation

In "Graphical estimation of slopes," you learned to estimate the slope of a graph by looking at it "close up," and observing that it resembles a straight line locally. Actually, the graphs you examined in that worksheet never really get perfectly straight. They just look that way from close-up. However, we can say that there are straight lines that approximate this part of the graph when you look at very small pieces. And among these approximating lines, there is one that is the best linear approximation to the graph near the point on which you are focusing your attention. Intuitively this means that no matter how close you get to the curve at this point, this "best" line still looks like a good approximation.

In this exercise, you will look at a close up view of a graph together with its best linear approximation at a point. Then you will "back off" and view the relationship of the entire line to the entire original graph.

For the first view, graph the functions $f(x) = x/(1 + x^2)$ and $L(x) = 0.48x - 0.16$, using a viewing window from about $x = -0.505$ to $x = -0.495$ and $y = -0.405$ to $y = -0.395$. The graph you see is a close up view of $f(x)$ near the point $(-0.5, -0.4)$, together with its best linear approximation, $L(x)$. Notice that you really see only one figure there. The graph of f and the graph of its best linear approximation appear to coincide (lie on top of each other) from this close in.

Now change the domain and range of the graph three times as follows (the given values are the new domain and range):

1. From $x = -0.55$ to $x = -0.45$ and from $y = -0.45$ to $y = -0.35$

2. From $x = -1$ to $x = 0$ and from $y = -0.9$ to $y = 0.1$

3. From $x = -3.5$ to $x = 2.5$ and from $y = -3.4$ to $y = 2.4$

From what you have observed, you should have a pretty good idea what the best linear approximation to a graph looks like. The best linear approximation to a graph at a given point is called the *tangent line to the graph at the point*. The slope of the tangent line, which we identified in the activity "Graphical Estimation of Slope" as the slope of the curve at the given point, is called the *derivative of the function at the given point*.

Here are the graphs we got.

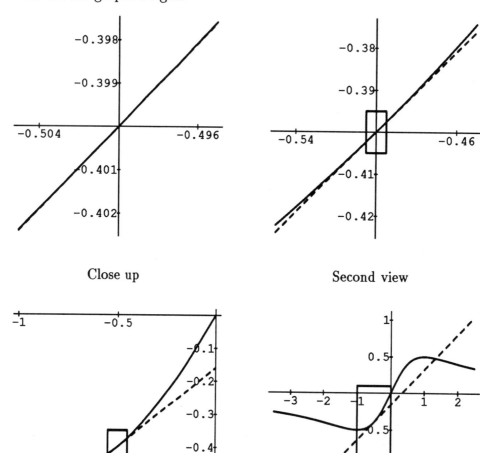

Close up

Second view

Third view

Last view

Have the students do another example.

Problem

1. "Slope with rulers" activity

Estimating cost

Topics: Derivative, linear approximation

Summary: This activity should help the students see that the marginal cost is the derivative of the cost function.

Background assumed: None required; if the class has already seen the idea of secant lines this activity can easily be related to the concept.

Time required: 10–15 minutes

Resources needed: None; a calculator is useful for the calculations.

Threads: approximation and estimation, multiple representation of functions

---

A clothing company makes dress shirts. The table below gives information about the total cost of making shirts on a typical day. The top line of the table gives the total number of shirts made since the beginning of the day and the bottom line gives the cost (in dollars) of making those shirts.

Number	0	10	20	30	40	50	60	70	80
Cost	1000	1316	1447	1548	1632	1707	1775	1837	1894

1. (a) What is the average cost of a shirt for the last 40 shirts that were made?
   (b) What is the average cost of a shirt for the last 20 shirts that were made?
   (c) What is the average cost of a shirt for the last 10 shirts that were made?

2. Based on the information you gathered in 1, do you think the graph of the cost near $(80, 1894)$ is increasing and concave up, increasing and concave down, decreasing and concave up, or decreasing and concave down? Explain your choice.

The curve seems to be increasing and concave down near (80, 1894) since the averages—the slopes of secant lines—are decreasing as the intervals over which the average is computed get smaller. This idea should be used to illustrate why the cost for one more shirt should be less than all the averages computed.

---

3. The company is trying to estimate how much it will cost to make one more shirt (in addition to the 80 already made).

    (a) Can the company make another shirt for less than $10? Explain.
    (b) Can the company make another shirt for less than $6? Explain.
    (c) What is your best estimate for the cost of one more shirt? Explain.

The average cost for the last 40 shirts is $6.55 per shirt; for the last 20 shirts the average cost is $5.95 per shirt and for the last 10 shirts the average cost is $5.70 per shirt. The best estimate should be less than $5.70 for the shirt.

Problems

Quarter horses race a distance of 440 yards (a quarter mile) in a straight line. During a race the following observations were made. The top line gives the time in seconds since the race began and the bottom line gives the distance (in yards) the horse has traveled from the starting line.

Time	2	4	6	8	10	12	14	16	18	20
Distance	30	64	100	138	177	217	259	302	347	397

1. What is your best estimate for how fast the horse is running halfway through the race?

2. The horse will win a bonus if the time for the race is less than 22 seconds. Decide whether you think the horse will win the bonus. Explain your reasons.

---

1. Since 12 seconds gives the information closest to halfway through the race (220 yards), use a difference near 12 seconds. Reasonable estimates are 40/2 = 20 yards/second or 42/2 = 21 yards per second. In fact, since the average velocities are increasing, the distance curve should be increasing and concave up at (12, 217), so the slope at 220 will be between 20 and 21 yards per second.

2. For the last 2 seconds, the average velocity is 50/2 = 25 yards per second and the horse has been speeding up throughout the race. So, we can expect the horse's velocity at the end of the race to be at least 25 yards per second. The horse must run 43 yards, and at 25 yards per second this will take 43/25 = 1.72 seconds. The horse should finish the race in less than 22 seconds. (Note that using the average for the first 20 seconds will result in an answer of 22.166 seconds for the race.)

Finite differences

Topics: Comparison of finite differences and derivatives

Summary: This activity shows that, under certain conditions, finite difference tables replicate information given by derivatives of polynomials; for example, the differences vanish at the same level as the derivative. The activity also has an "antiderivative" component.

Background assumed: Difference quotient

Time required: 15 minutes

Thread: approximation and estimation

Part A

1. Each person in the group should choose a different polynomial (preferably of differing degrees ≤ 3) and complete the table below as follows. If your polynomial is $y = p(x)$, in Row 1 compute $p(1)$, $p(2)$, $p(3)$, $p(4)$, $p(5)$, $p(6)$. In Row 2, over the star "*" compute $(p(2)-p(1))$, $(p(3)-p(2))$, $(p(4)-p(3))$, $(p(5)-p(4))$, and $(p(6)-p(5))$. In Row 3 and in subsequent rows, over the stars, compute the difference between the entry in the row directly above the star to the right and the entry in the row directly above the star to the left. See Example 1.

1	2	3	4	5	6	Row 0
*	*	*	*	*	*	Row 1
	*	*	*	*	*	Row 2
		*	*	*	*	Row 3
			*	*	*	Row 4
				*	*	Row 5
					*	Row 6

Example 1: $f(x) = x^2$

1		2		3		4		5		6		Row 0
1		4		9		16		25		36		Row 1
	3		5		7		9		11			Row 2
		2		2		2		2				Row 3
			0		0		0					Row 4
				0		0						Row 5
					0							Row 6

2. As a group determine the similarities in the tables. What general conclusions—for all polynomials, at least—would you conjecture?

Part B

In part A, Row 2 contains numbers of the form $p(n+1) - p(n)$. Such a number can also be written as $\frac{(p(n+1) - p(n))}{1}$ or $\frac{(p(n+h) - p(n))}{h}$ where $h = 1$. Thus, this number is an approximation to $p'(n)$. Similarly, Row 3 contains approximations of p'', etc. (Notice in the example that the values in Row 3 *are* the values of the second derivative of $y = x^2$.) We now want to see if we can "recover" a function if we know something about its "finite difference table"; that is, we want to anti-differentiate from a table. We will concentrate on determining quadratic functions.

1. Each person in the group should write in a different number in the right-most position in Row 1 and then complete the table so that the table corresponds to a table for a function as in part A. Now, try to determine the function.

1		2		3		4		5	6	Row 0
*		*		*		*		*	*	Row 1
	*		*		*		*		1	Row 2
		2		2		2		2		Row 3

As a group, decide what the first entry you chose for Row 1 does? Could you put more than one entry in Row 1 to start and then complete the table?

2. As individuals, take the table in 1 and put a different number in the right-most entry of Row 2 and choose a number for the right-most entry in Row 1 as before and complete the table. Can you write an algebraic expression for the function? As a group, determine what role an initial entry in Row 1 plays in the expression for the function.

Problems

Consider a table for a quadratic function that begins as follows:

1		2		3		4		5	6	Row 0
*		*		*		*		*	*	Row 1
	*		*		*		*		*	Row 2
		1		1		1		1		Row 3

Determine at least one quadratic function that has such a table. What role does the value in Row 3 play in the determination of the algebraic expression of the function?

Using the derivative

Topics: Increasing, decreasing functions; concavity; local maximum

Summary: Students discover the relationship between the properties of the graph of a function (increasing, decreasing, concave up, concave down) and the derivative.

Background assumed: None

Time required: 15 minutes

Thread: graphical calculus

Below are graphs of a function f and its derivative f'.

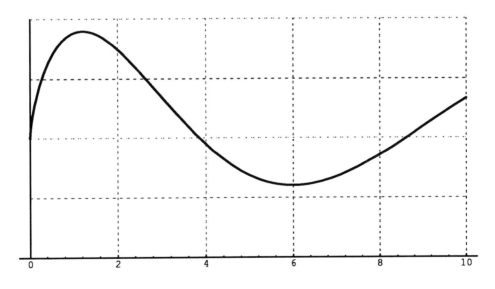

$y = f(x)$

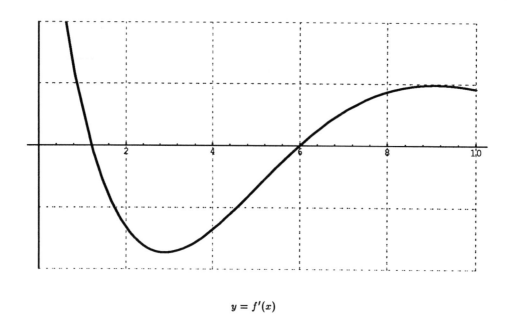

$y = f'(x)$

1. For what values of x is f increasing?

2. What is the behavior of f' on those intervals?

3. For what values of x is f' negative?

4. What is the behavior of f on those intervals?

5. For what values of x is f' increasing?

6. What is the behavior of f on those intervals?

7. Where is the graph of f concave down?

8. What is the behavior of f' on those intervals?

9. Where does f take on a local maximum value?

10. What can you say about f' there?

Students should see that f is increasing [resp., decreasing] where f' is positive [negative] and f is concave up [down] where f'' is positive [negative].

A function f has a local maximum where f stops increasing and starts decreasing; that is, f' changes from positive to negative.

Ask students to characterize a local minimum in terms of the derivative and to verify their conjecture by consulting the local minimum on the graph.

Gotcha

Topic: Mean value theorem

Summary: Students see an application of the mean value theorem.

Background assumed: Derivatives

Time required: 15 minutes

Threads: modeling, distance and velocity, graphical calculus

The following passage comes from the article " 'I think I'll go down to Tully' and catch some speeders," in the Sunday, June 20, 1993, Syracuse *Herald American*.

> Trooper Jay Schwenk sat behind the pilot in the state police plane and surveyed the traffic 2,000 feet below.
> Click. Schwenk started the sky timer as a car zoomed over a white stripe painted on Interstate 81 in Tully. Click. Schwenk stopped the timer when the car hit a second stripe, a quarter mile from the first.
> "Eighty-three on the white one," Schwenk told the pilot, Sgt. Gerd Wolf. Schwenk timed the car again just to be sure. This time the driver was clocked at 82 mph.
> "I've got one coming up now going 82," Schwenk told patrol cars waiting alongside the highway just south of the Onondaga County line. "It'll be a white vehicle in the passing lane."
> Seconds later, a trooper pulled the car over.
> ...

As a group, answer the following.

1. How do you think that the car speed was computed?

2. Do you think that the driver had to be actually traveling the reported car speed at any instant?

3. Draw at least two different distance functions and their associated velocity functions for situations where the quarter mile would have been traversed in 10 seconds. What would have the trooper report? In each case is there a time when the velocity equals this reported speed?

Problem

Gas is flowing out of a pipe in such a way that it fills a one-gallon container in 10 seconds. Do you think it is possible that the rate at which the gas is flowing into the container is never 0.1 gallons per second? If so, construct an example of a function that gives the amount of gas in the container as a function of time where the rate of flow is never 0.1 gallons per second. Give a concrete scenario for your function. If you don't think this is possible, justify your conclusion.

The newspaper article shows that this situation is not contrived, and most students can relate to such a plausible scenario. The homework is meant to stress the necessary condition that the function be differentiable over the interior of the time interval in question. A rupture in a pipe could account for non-differentiability, if not discontinuity, which could lead to a case where the rate is never equal to the average flow.

Animal growth rates

Topics: Differential equations, differentials

Summary: Students are asked to think qualitatively about a differential equation and to use the tangent line to approximate the value of a function based on its value at a point and its derivative.

Background assumed: Rates

Time required: 20 minutes

Threads: approximation and estimation, graphical calculus

You are involved in a research project that involves working with a species of laboratory animal. If $W(t)$ is the weight (in ounces) of such an animal t weeks after birth, then the growth of a healthy animal can be modeled by the differential equation

$$W'(t) = \frac{10}{W(t)} \quad \text{or} \quad \frac{dW}{dt} = \frac{10}{W}.$$

1. Describe in words what the differential equation says about the growth of a healthy animal.

Suppose you are responsible for an animal that weighs 5 ounces one week after it was born.

2. What does the model tell us about the rate of growth of the animal when it is one week old?

3. Give the equation of the line tangent to the graph of $W(t)$ at the point $(1,5)$.

4. Use the tangent line to approximate the weight of the animal 8 days after it was born.

   _____

   Be careful with units here.

   _____

5. Classify the graph of $W(t)$ for $t \geq 1$ as increasing and concave up, increasing and concave down, decreasing and concave up, or decreasing and concave down.

Problem

Suppose instead that the growth is modeled by

$$W'(t) = \frac{10}{W(t)^2} \quad \text{or} \quad \frac{dW}{dt} = \frac{10}{W^2}.$$

Answer 1–5 above.

The product fund

Topics: Product rule

Summary: Students derive the product rule.

Background assumed: Definition of derivative

Time required: 40 minutes

Thread: top-down methodology

_____

Pat is employed by a financial services firm. One of her tasks is to execute transactions for her best client. The client is currently investing in the Drake mutual fund. The client has decided to purchase shares in the fund everyday instead of simply making a single large purchase. (This method of investing is sometimes called an "averaging" method since buying at several different times will mean that the price paid for the mutual fund will tend to be about the average price of the mutual fund over the time period. If the client purchased all her shares at once the client runs the risk of buying at the highest price which occurs during a period of time.)

1. The client owns 1000 shares of the Drake Fund at the start of business on Monday morning. The client calls Pat and wants to know the value of her Drake Fund investment. Pat looks at her workstation and sees that Drake Fund is currently trading at $13.50 per share. What is the current value of the client's Drake Fund investment?

2. By the end of the trading day on Monday two things have happened: (1) the price of a share of the Drake Fund has fallen $0.25, and (2) Pat has purchased 100 more shares of the Drake Fund for her client.

    (a) What is the value of the client's investment in the Drake Fund at the end of the day?
    (b) What was the change in the value of the investment during the day?

Investors are very interested in whether the value of their investment is growing (this makes them happy) or declining (this makes them unhappy). However, investors are also very interested in the rate of change of the value of their investment. An investment growing at the rate of $10 per day is much better than an investment that is growing at the rate of $10 per year.

3. Pat decides to call the value of the investment of her client in the Drake Fund t days after the start of business on Monday morning $V(t)$. She will call the number of shares of the Drake Fund that her client owns t days after the start of business on Monday morning $N(t)$. She will denote by $p(t)$ the price of a share of the Drake Fund t days after the start of business on Monday morning.

    (a) At any time t there is a formula that relates V, N, and p. Write this formula.
    (b) Write the information contained in 1 in terms of V, N, p, and t.
    (c) Write a sentence that describes what $V'(t)$ means in the context of this activity.
    (d) What are the appropriate units for $V'(t)$?
    (e) Write a sentence that describes what $N'(t)$ means in the context of this activity.
    (f) What are the appropriate units for $N'(t)$?
    (g) Write a sentence that describes what $p'(t)$ means in the context of this activity.
    (h) What are the appropriate units for $p'(t)$?

4. Pat's client wants information about $V'(t)$, the rate of change of her investment. Pat has information about $N'(t)$, the rate of change of the number of shares that her client owns, since she has her client's purchase orders. She can look in the newspaper or use her workstation and get information about $p'(t)$, the rate of change of the price of a share of the Drake Fund. So, like any problem solver, she has to figure out how to use what she knows to get what she wants.

We will try to figure out what $V'(t)$ is by breaking the problem into two separate parts. (A good strategy for problem solving is to break a big problem into smaller problems and then try to solve the smaller problems. This is part of the *top-down* approach to problem solving.)

We will break the value of the investment into two parts: the first part is the value from shares bought at times less than or equal to t, and the second part is the value from shares bought *after* time t. We will call the shares bought at times less than or equal to t "old shares." We will call the shares bought after time t "new shares." Any share is an old share or a new share but *not both*, so the value of the investment is the total of the value of the new shares and the value of the old shares.

We will estimate what happens to the value of the client's investment dt days after t. Using our strategy from above we will use the two categories of shares, "old shares," which were owned by time t, and "new shares," which were purchased some time after time t but no later than time $t + dt$.

(a) Write down an expression for the number of old shares using N.

(b) Write down an expression for the number of new shares using N.

You should have obtained the following results: the number of old shares is $N(t)$ and the number of new shares is $N(t+dt) - N(t)$.

(c) Write a formula or expression for the value of the old shares at time $t + dt$.

(d) Write a formula or expression for the change in the value of the old shares between time t and time $t + dt$.

(e) Since we are interested in the rate of change, divide the change in value of the old shares by dt and let dt approach 0. You now have the rate of change in the value of the old shares at time t.

(f) Write a formula or expression for the value of the new shares at time $t + dt$.

(g) How many new shares had been purchased by time t?

(h) What was the value of the new shares at time t?

Be sure that the class understands that the number of new shares at time t is 0. Thus the change in value of the new shares will be $(N(t+dt) - N(t))(p(t+dt) - 0)$.

(i) Write an expression for the change in the value of the new shares between time t and time $t + dt$.

(j) What mathematical property of the function p will guarantee that $p(t+dt) \to p(t)$ as $dt \to 0$?

(k) Since we are interested in the rate of change, divide the change in value of the new shares by dt and let dt approach 0. You now have the rate of change in the value of the new shares at time t.

(l) Add the expressions you obtained in parts 4e and 4k to obtain an expression for the rate of change in the value of the investment at time t.

(m) Write a sentence that will tell Pat how to calculate the rate of change of the value of the investment based upon what she knows.

5. Summarize what you have learned about finding the derivative of any function V which is written as the product of two other functions such as N and p in terms of the functions N and p and their derivatives.

Problems

1. The price of a share is $14 and is rising at the rate of $0.10 per day. Pat's client has 700 shares of the mutual fund and has instructed Pat to buy shares at the rate of 100 per day. What is the rate of change of the value of the client's investment in the mutual fund?

2. Pat knows that the shares in the mutual fund can't change by more than $1.50 each day. (The exchange will halt trading in the shares if this happens.) She also has received instructions from her client to buy between 100 and 200 shares in the mutual fund each day. Pat is trying to get some idea of the rate of change of the value of the investment in the Drake fund. At the beginning of the day her client has 800 shares and each share is currently worth $13.

   (a) Draw a graph of the price of a share for a day that will force the exchange to halt trading.

   (b) Draw a graph of the price of a share for a day that will *not* force the exchange to halt trading.

(c) Decide whether it is possible to draw a graph of the price of a share for a day which will *not* force the exchange to halt trading but will have at least one point on the graph where the derivative of the price is larger than $1.50/day.

(d) Write a brief paragraph explaining why or why not Pat can give her client upper and lower estimates for the rate of change of the value of her investment at the beginning of the day.

(e) Write a brief paragraph explaining why or why not Pat can give her client upper and lower estimates for the value of her investment at the end of the day.

When you summarize this activity, a geometric explanation is very helpful. Draw two coordinate axes and label one $N(t)$ and the other $p(t)$. Now $V(t)$ can be thought of as the area of a rectangle with sides $N(t)$ and $p(t)$. The change in V corresponds to the area of an L-shaped region. This L-shaped region is the difference between a rectangle with sides $N(t+dt)$ and $p(t+dt)$ and a rectangle whose sides are $N(t)$ and $p(t)$. The L-shaped region can be divided into three rectangles. One rectangle corresponds to the change in the value of the old shares, and a second rectangle corresponds to the change in the value of the new shares. The third is "small."

Exchange rates and the quotient rule

Topic: Quotient rule

Summary: Students investigate difference quotients numerically in the case $g = 1/f$ to verify the quotient rule.

Background assumed: Difference quotients

Time required: 30 minutes

Resources needed: Calculator

Thread: approximation and estimation

Recall that
$$\frac{d}{dx}\frac{1}{f(x)} = -\frac{1}{f(x)^2}f'(x).$$
Here we will investigate this phenomenon in the context of currency exchange rates.

According to the *New York Times*, the exchange rate for the Japanese yen on Friday, July 16, 1993, was 107.55 yen per U.S. dollar. The *Times* also reports that on the previous Friday the exhange rate was 109.94 yen per dollar.

Let $f(x)$ denote the exchange rate for yen (measured in dollars). Let $g(x)$ denote the exchange rate for dollars (measured in yen). For purposes of this activity, we will assume that $g(x) = \frac{1}{f(x)}$. (Strictly speaking, this is not accurate. At any time, if you exchange yen for dollars and then dollars for yen (or *vice versa*), you will end up with less money than you started.)

1. (a) Assume that you are holding dollars and you want to shop using yen. Did your dollars gain or lose value this week?

   (b) Assume that you are holding yen and you want to shop using dollars. Did your yen gain or lose value this week?

Dollars lost value; yen gained value.

---

2. What do your answers to 1a and 1b say about the signs of $f'(x)$ and $g'(x)$?

3. Units are very important in this situation.

   (a) What are the appropriate units for $f(x)$?

   (b) What are the appropriate units for $g(x)$?

Units for f are $\frac{\text{yen}}{\text{dollar}}$; for g, $\frac{\text{dollar}}{\text{yen}}$.

---

4. Using the data above, estimate f' on Friday. (Use weeks to measure time. What are the units for f'?)

$$f'(x) \approx \frac{107.55 - 109.94}{1} = -2.39 \frac{\text{yen}}{\text{dollar}}}{\text{week}}.$$

5. Using the data above, estimate g' on Friday. (What are the units for g'?)

$$g'(x) \approx \frac{\frac{1}{107.55} - \frac{1}{109.94}}{1} = 0.000202 \frac{\frac{\text{dollar}}{\text{yen}}}{\text{week}}.$$

6. Check the quotient rule. Do the units work out? Can you explain why the answers do not match exactly?

Units match, but the numbers don't (since a non-constant function and its reciprocal can't both be linear).

Problems

1. Use your estimate from 4 to estimate the exchange rates ($f(x)$ and $g(x)$) on Thursday, July 15.

2. Using this data point, estimate g' on Friday. How does it compare to what you found in 5?

Using the product rule to get the chain rule

Topics: Derivatives

Summary: Students obtain the special case of the chain rule that gives the derivative of a power of a function.

Background assumed: The product rule for derivatives; for the second part, the quotient rule for derivatives.

Time required: 10 minutes for the first part; 5 minutes for the second

---

Part A

When you learned operations such as multiplication you only learned results such as $5 \times 3 = 15$. If you needed to compute the product of $3 \times 5 \times 4$ you used the idea of *associativity*. You computed $3 \times 5 = 15$ and then multiplied $15 \times 4 = 60$. This could be denoted $(3 \times 5) \times 4$.

1. Since you know how to find the derivative of a product of two functions, use the idea of associativity to find the derivative of the product of three functions f, g, and h. Your answer should involve f, f', g, g', h, and h'.

2. Find a rule for the derivative of the product of four functions f, g, h, and k.

3. Use the product rule to find the derivative of the product of two copies of the same function f. (We will denote this product as f^2.)

4. Use your result from 1 to find the derivative of the product $f \times f \times f$ or f^3.

5. Use your result from 2 to find the derivative of the product $f \times f \times f \times f$ or f^4.

6. What do you think the derivative of f^n is for any positive integer n?

Part B

1. Use the quotient rule to find the derivative of $1/f$. Using the traditional exponential notation we will call this f^{-1}. *Note: This does not stand for the inverse of f.*

2. Use the product rule to find the derivative of $1/f \times 1/f$. (We will write f^{-2} to denote $1/f \times 1/f$.)

3. Write f^{-3} as $1/f \times 1/f \times 1/f$ and use your result from 1 of part A to find the derivative of f^{-3}.

4. Write f^{-4} as $1/f \times 1/f \times 1/f \times 1/f$ and use your result from 1 of part A to find the derivative of f^{-4}.

5. What do you think the derivative of f^n is for any negative integer n?

6. What do you think the derivative of f^n is for any integer n?

Problems

1. Compute the derivative of $\sin^{12} x$.

2. Is there a function whose derivative is $\sin^{11} x \cos x$?

3. (a) Find the derivative of e^{nx} for any integer n.
   (b) What do you think the derivative of e^{kx} is for any number k?

Magnification

Topic: Derivatives

Summary: Students use the idea of a function as a lens to develop the topic of magnification and its relationship to the derivative.

Background assumed: Difference quotients

Time required: 30 minutes

Thread: modeling

The *image* of an interval $[a,b]$ under a map f (this is denoted as $f([a,b])$) is the set of all values y that are equal to $f(x)$ for some x in $[a,b]$. In shorthand,

$$f([a,b]) = \{y : y = f(x) \text{ for } a \leq x \leq b\}.$$

For example if $f(x) = x/2$, then $f([-2,2]) = [-1,1]$; if $f(x) = x^2$, then $f([-2,2]) = [0,4]$.

For each of the following functions find the image of each interval, the length of the image, and the length of the image divided by the length of the original interval. What do you think the magnification due to f is when $x = 3$ for each of these functions?

1. If $f(x) = 3x$

$[a,b]$	length	$f([a,b])$	length	ratio
$[2,4]$	$4-2=2$			
$[2.5, 3.5]$	1			
$[2.9, 3.1]$	0.2			
$[2.99, 3.01]$	0.02			
$[2.999, 3.001]$	0.002			

Magnification at $x = 3$ is _____ .

2. If $f(x) = x^2$

$[a,b]$	length	$f([a,b])$	length	ratio
$[2,4]$	2			
$[2.5, 3.5]$	1			
$[2.9, 3.1]$	0.2			
$[2.99, 3.01]$	0.02			
$[2.999, 3.001]$	0.002			

Magnification at $x = 3$ is _____ .

3. If $f(x) = -x^3$

$[a,b]$	length	$f([a,b])$	length	ratio
$[2,4]$	2			
$[2.5, 3.5]$	1			
$[2.9, 3.1]$	0.2			
$[2.99, 3.01]$	0.02			
$[2.999, 3.001]$	0.002			

Magnification at $x = 3$ is _____ .

4. If $f(x) = 1/x$

$[a,b]$	length	$f([a,b])$	length	ratio
$[2,4]$	2			
$[2.5, 3.5]$	1			
$[2.9, 3.1]$	0.2			
$[2.99, 3.01]$	0.02			
$[2.999, 3.001]$	0.002			

Magnification at $x = 3$ is _____ .

5. What do you think the magnification means *mathematically*? Include any other conjectures you think are true about magnification.

The students should see that $|f'(3)|$ is the magnification and that the sign gives the orientation. Orientation is worth mentioning.

You can justify the chain rule geometrically very nicely after this activity. The composition of two functions corresponds to applying two lenses (one corresponding to each function), and the magnification of the composition is the product of the magnification of each.

Chapter 4

Integration

Time and speed

Topic: Integration

Summary: Introduces Riemann sums as estimates for the integral. Since the function is monotone, upper and lower estimates for the distance can be obtained.

Time required: 10–15 minutes

Threads: distance and velocity, approximation and estimation, multiple representation of functions

---

A car moved along a straight road and its speed was continually increasing. Speedometer readings were recorded at two-second intervals and the results were as follows:

Time	0	2	4	6	8	10
Speed	30	36	38	40	44	50

The speeds are given in feet per second, the times in seconds.

1. From the above information, one cannot tell exactly how far the car went in the ten seconds. Explain why this is true.

2. In the first two-second interval, what is the minimum distance the car could have traveled? What is the maximum distance?

3. In the second two-second interval, (i.e. between $t = 2$ and $t = 4$) what is the minimum distance the car could have traveled? What is the maximum distance?

4. In the third two-second interval, what is the minimum distance the car could have traveled? What is the maximum distance?

5. In the fourth two-second interval, what is the minimum distance the car could have traveled? What is the maximum distance?

6. In the fifth two-second interval, what is the minimum distance the car could have traveled? What is the maximum distance?

7. During the entire ten-second interval, what is the minimum distance the car could have traveled? What is the maximum distance? (These distances are called the *lower estimate* and the *upper estimate* of the distance traveled, respectively.) Explain where you used the assumption that the car's speed was increasing. What assumption could replace it and still give the same answer? What would the ride be like under your assumption?

8. If you had to guess how far the car went in the ten-second interval, what would you guess? What is the maximum difference (error) between your guess and the actual distance?

Explain that in this case the average of the two estimates will minimize the worst possible error.

The problems give error estimates for the integral of a monotone function, either by telescoping sums or by a geometric argument.

Problems

Suppose the following additional speedometer readings become available at the missing one-second times:

Time	1	3	5	7	9
Speed	32	37	39	41	49

1. Is the second set of information consistent with the first? What would the speed at the time 1 second have to be for it to be inconsistent?

2. Redo Questions 7 and 8 in the light of this new information.

3. If speedometer readings became available for each tenth of a second, i.e. for $t = 0.1, 0.2, 0.3, \ldots, 9.8, 9.9$, then by how much would your upper estimate for the distance traveled exceed your lower estimate?

4. If speedometer readings were available for each hundredth of a second, by how much would your upper estimate for the distance traveled exceed your lower estimate?

5. Explain why you can calculate how far the car went to any desired accuracy if you have access to the speedometer readings at every instant during the ten-second interval.

Note that
$$(\text{upper estimate} - \text{lower estimate}) \leq (f(10) - f(0)) \cdot h$$
where h is the difference between successive time measurements.

Oil flow

Topics: Rates, relationship of area under the rate curve and the total change in the function

Summary: Given the rate of flow of oil in a spill for a particular time period, students estimate the total amount of oil spilled during that time period. The flow rate is given as a table of values.

Background assumed: Relationship between velocity and distance (in particular, the relationship between f' and f)

Time required: 30 minutes

Threads: graphical calculus, modeling, approximation and estimation

---

Earlier this week, an oil tanker collided with a Coast Guard cutter off the California coast. The disabled tanker is spilling oil from its damaged hull.

The rate of flow of oil into the Pacific Ocean off the California coast was measured at several different time intervals yesterday. The rates are listed in the table below.

Time	Amount (100 gal/hour)
9:00	4.0
10:00	4.0
11:00	3.8
12:00	3.6
1:00	3.0
2:00	2.0
3:00	0.6
4:00	0.3
5:00	0.1
6:00	0.1
7:00	0.0

1. Draw a graph of the rate at which oil was spilling into the ocean as a function of time.

2. Estimate the total amount of oil that spilled during the ten hour period covered by the table. Explain your method. That is, explain any assumptions you are making.

3. What can you say about the error involved in part 2?

4. What extra information would allow you to improve your estimate in part 2? Try to be as specific as you can.

Comment: This activity—along with the activities "Time and speed" and "Can the car stop in time?"—gives a good introduction to definite integration and numerical approximation as a method of calculating area. In particular, it seems reasonable from the data given that the oil flow rates are decreasing, so upper and lower estimates can be used to bound the total amount of oil spilled. Make sure the students pay particular attention to the fact that the table gives rates of flow.

Can the car stop in time?

Topic: Integration

Summary: Students use an upper estimate for an integral in order to solve a problem.

Time required: 5–10 minutes

Threads: distance and velocity, multiple representation of functions, approximation and estimation

Part A

A car is traveling at 60 feet per second when the driver spots a deer in the road 300 feet ahead and slams on the brakes. The following readings of the car's speed at various times are given in the table below. Time is measured in seconds (since the driver applied the brakes) and speeds are measured in feet per second.

Time	0	2	4	6	7
Speed	60	50	30	12	0

1. Give an estimate for the minimum distance the car will travel after the brakes are applied. Show your work. State any assumptions you are making.

2. Give an estimate for the maximum distance the car will travel after the brakes are applied. Show your work.

The last time interval is only one second, not two seconds, so the maximum distance for that time interval is 12 feet, not 24 feet.

---

3. If the deer freezes and does not move, will the car hit the deer? Explain.

Part B

If the car was traveling at 60 feet per second when the driver saw the deer and the car slowed down at the rate of 8 feet per second per second, would the car hit the deer? Explain.

This question can be answered by computing a Riemann sum, by integrating $v(t) = 60 - 8t$, or by finding the area under the velocity curve. Working this problem each of the three ways is a good review of the different interpretations of integration for functions represented in different ways.

Fundamental theorem of calculus

Topic: Fundamental theorem of calculus

Summary: Students discover a proof of the fundamental theorem of calculus for the special case of a linear function.

Background assumed: Area of triangles and trapezoids, function notation, derivatives

Time required: 35–40 minutes

Thread: multiple representation of functions

In this activity we will be trying to find areas under curves. We will start with a simple example.

1. Sketch the curve $y = 2t + 3$.
2. Find the area under this curve between the lines $t = 1$, $t = 4$, and the t-axis using geometry.
3. Find the area under this curve between the lines $t = 1$, $t = x$, and the t-axis using geometry.

Your answer to question 3 should involve x, and we can think of this formula as representing a function which gives the area under the curve between $t = 1$ and $t = x$. We will call this area function $A(x)$.

4. Evaluate $A(4)$ and check that you get the same answer as you did to question 2.

Next we will look at the rate of change of A with respect to x.

5. Pick a value $x > 1$, and indicate the region on your graph from question 1 whose area corresponds to $A(x)$.

When students get to the definition of $A(x)$, the pictures asked for in question 5 help them understand the concept. If one student in a group draws $A(2)$, another $A(3)$, and so on, this will help the students get the idea of the new function.

---

6. Pick a value of $h > 0$ and indicate the region on your graph from question 1 whose area corresponds to $A(x + h)$.

7. Use geometry to find an algebraic expression for the area of the region in your original graph which corresponds to $A(x + h) - A(x)$. (Your answer will involve both x and h.)

8. Divide your answer to question 7 by h.

9. Look at your answer to question 8 and let $h \to 0$.

10. In the language of calculus describe what questions 7, 8, and 9 accomplished.

11. Look at your answer to question 9 and the original function defined in question 1. How are they related?

12. Describe in the language of calculus how the function A you found in question 3 and the original function given in question 1 are related.

Problem
If the function you were given was not linear (such as $x^3 + 2$)—so that you could not find $A(x)$ by geometry—can you think of another way to find $A(x)$ that will work? Write a detailed description of your method, using the language and notation of calculus.

Comparing integrals and series

Topics: Riemann sums, area as a sum

Summary: This activity is intended to: focus on series as a limit; emphasize the similarities and differences between Riemann sums and series early in a discussion of series; provide a geometric representation of a series; point out that a partial sum of such a series can be "relined" to form a Riemann sum; and prepare students for the graphical arguments usually given for the integral test for series.

Background assumed: Representation of an integral as the limit of Riemann sums; integral as area; closed formula for geometric series

Time required: 35 minutes

Thread: approximation and estimation

---

1. Figure 1 (below) indicates a pattern for the creation of infinitely many boxes inside a triangle in the plane. If the areas of all the boxes were collected to form a series, think about whether or not the sum of the series would be the area of the triangle. Suppose your group was a group of attorneys and that your firm had to defend an answer to this question in a courtroom, with this figure, in evidence and where the jurors had not had calculus. You will need to consider arguments for both a yes answer and a no answer to this question. Without computing the sum of the series at this point, have your group prepare an intuitive argument why the answer might be yes as well as an intuitive argument why the answer might be no. Which argument does your group actually support at this time?

Discussion of 1 should take place before the rest of the activity is assigned. Some groups may conclude that it is obvious that the series converges to the area of the triangle and, thus, be unwilling to try to provide a counter argument.

One way to handle this is to ask: As successive "rows" of boxes are piled up, to what length does the "skyline" converge? (The skyline is the path traced out using the tops of the highest boxes joined together by portions of the sides of appropriate boxes.) This skyline converges to the hypotenuse of the triangle. So, perhaps the skyline length should approach the length of the hypotenuse. But the sum of lengths of the vertical pieces and horizontal pieces of successive skylines can be seen to converge to the sum of the lengths of the two sides of the triangle. (Project the horizontal pieces onto the x-axis and the vertical pieces onto the y-axis.) These two observations give different results! So, the result may *not* be so obvious. (This latter example is also useful when discussing arc length. It suggests why the vertices of piecewise approximations to the arc must lie on the arc.)

---

2. *Compute* the actual sum of the series from part 1. Does the sum of the series equal the area of the triangle? Does this provide conclusive evidence for either a yes answer or a no answer to the question raised in part 1?

The pattern is
$$\left(\frac{1}{2}\right)^2 + 2\left(\frac{1}{4}\right)^2 + 4\left(\frac{1}{8}\right)^2 + \cdots;$$
that is,

    area of biggest box + area of next bigger boxes + area of $\cdots$

The series is
$$\sum_{n=0}^{\infty} 2^n \left(\frac{1}{2^{n+1}}\right)^2 = \sum_{n=0}^{\infty} \frac{1}{4}\left(\frac{1}{2}\right)^n,$$
a geometric series with sum
$$\frac{1}{4(1-\frac{1}{2})} = \frac{1}{2}.$$

The area of the triangle is
$$\frac{1}{2}bh = \frac{1}{2} \times 1 \times 1 = \frac{1}{2}.$$

So, mathematically, this proves that the area of the triangle is the sum of the areas of the squares. However, it doesn't mean that the jury will believe this answer.

---

3. *Represent* the area of the triangle as an integral.

The hypotenuse is the graph of $y = 1 - x$ over the interval $[0, 1]$. So, the area is $\int_0^1 (1-x)\,dx$. Students will want to compute the value of the integral.

---

4. What is the relationship between the value of the integral and the sum of the series? Does the series represent a Riemann sum?

The value of the integral and the sum of the series are the same. However, the series does not represent a Riemann sum. (A Riemann sum is a finite sum.)

5. Each member of your group should draw a (different) graph of a continuous, positive valued function, $y = f(x)$, over a closed interval in the plane. As a group, do you think that $\int f(x)\,dx$ can be represented as a series for any such function f? Support your conclusions.

Yes, every such integral can be represented by a series. Each partial sum of the series corresponds to a Riemann sum over a partition of the interval. (One way of doing this is by taking lower Riemann sums and choosing the partitions so that each is a refinement of the previous one.) The Riemann sums converge to the integral, and this is the limit of the partial sums of the series. So, the values of the integral and the series are the same. Some people indicated that the series would only represent an integral if the series could be expressed algebraically. (This is precisely the notion that this question seeks to change.)

Problems

1. For $m > 0$, determine a series that gives the area under the curve $y = 1 - mx$ over the interval $[0, \frac{1}{m}]$.

One series is $\sum \frac{1}{4m}(\frac{1}{2})^n$. Others can be formed using other partition points.

2. Formulate a possible "real life" scenario around incremental depletion of a resource or commodity, such as a certain percent of the total resource or commodity was depleted each year, where the question as to whether or not such incremental depletion would exhaust the resource in the long run might be argued in a courtroom. Modify the arguments your group, or some group, gave in 1 of the activity to fit this scenario.

Use any finite natural resource or the ozone layer for an environmental application. A financial reserve makes a good economic example.

---

3. Using software (e.g., *Mathematica*), write a routine that "generates a series" solution to the area under a curve determined by the graph of a positive-valued, decreasing, concave down function over a closed interval. That is, find an algorithm that builds a partial sum approximation to the area. Discuss advantages and disadvantages of this algorithm over the right-hand endpoint algorithm.

One difference between a series algorithm and an endpoint algorithm is that with a series algorithm, the $n+1^{st}$ sum, which would be a better estimate, is obtained from the information needed to compute the nth sum. That is, one would be able to use a recursive algorithm that continued to run until the absolute value of the difference between two subsequent partial sums was less than a preassigned positive number (with appropriate safeguards). Typically, in an endpoint algorithm, a new estimate of the integral would need to be computed essentially from scratch. On the other hand, a series algorithm might require more memory.

Graphical integration

Topic: Integration

Summary: Students estimate integrals graphically by counting grid boxes.

Time required: 25 minutes

Threads: graphical calculus, multiple representation of functions, approximation and estimation

---

1. Estimate the area of the region in the graph between $x = 1$, $x = 5$, $y = f(x)$, and the x-axis.

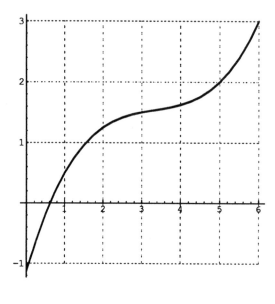

$y = f(x)$

Using the grid shown and counting blocks, we see that the area is between 3 and 8. We can obtain a more accurate approximation by dividing the x-axis and y-axis into units of one half (dividing each block into four blocks). These are the blocks we refer to below.

The area is between 19 and 27 blocks and a good estimate is about 22 blocks. Since each block has area $1/2 \times 1/2 = 1/4$ the area is about 5.5. If some students in the class are not familiar with estimating answers, this is a good opportunity to introduce the concept of good estimates and poor estimates. Using the idea of counting blocks should enable the students to find an upper and lower estimate easily. You should also point out that if the graph had smaller blocks you would expect to be able to get more accurate estimates. This idea can be related to Riemann sums. As the width of the blocks approaches 0, the upper and lower estimates will approach the actual area.

---

2. Find the volume of the region formed by revolving the region described in 1 around the x-axis.

The volume is about $(265/32)\pi = 26.0163\ldots$. This is a good time to point out that the area squared times π is *not* the same as the volume.

3. Find the area between $y = f(x)$, $y = 3$, $x = 1$, and $y = 1.5$.

The area is between 21 and 27 blocks (5.25 and 6.75) and a good estimate is 23.5 blocks or about 5.875 square units.

4. Find the volume formed by revolving the region described in 3 around the y-axis.

Using midpoints of washers that are $1/2$ wide gives approximately 38.9375π or about 122.326 cubic units.

Problems

1. Find the average value of the function $f(x)$ over the interval $[1, 5]$.

2. (a) Find the area of the region between $x = 1.5$, $y = f(x)$, $x = 3$, and the x-axis.
   (b) Express the area in 2a as a definite integral.

3. Find the volume of the solid formed by revolving the region described in 2a around the x-axis.

4. Find the area of the region between $y = 2.5$, $y = f(x)$, $y = 1.5$, and $x = 2$.

5. Find the volume obtained by revolving the region described in 4 around the y-axis.

6. How far is it along the curve $y = f(x)$ from $(1, 0.5)$ to $(3, 1.5)$?

Our intention was to have the class approximate the curve by straight line segments and measure each line segment. However, some students used a hair or a thread to follow the curve and then straightened and measured the length. In this case the idea of scale can be brought into the class. This problem also is a good introduction to arc length.

How big can an integral be?

Topics: Integration

Summary: This illustrates that several curves through the same set of data points can have very different integrals.

Time required: 30 minutes

Threads: graphical calculus, multiple representation of functions, modeling

Parts B and C can be omitted, if desired. See also "Projected image."

Part A

1. Plot the points

   $(0,1)$, $(1,2)$, $(2,3)$, $(3,4)$, and $(4,5)$.

2. Draw a straight line that passes through these points and is defined for all values in $[0,5]$.

3. Calculate the area under the graph you drew in 2, from $x = 0$ to $x = 5$.

4. Express your answer to 3 as an integral.

5. Draw a continuous curve that passes through the points and whose integral over $[0,5]$ is greater than 25.

6. Draw a continuous curve that passes through the points and whose integral is greater than 125.

7. Draw a continuous curve that passes through the points and whose integral is greater than 500.

8. Is there a positive number that is larger than the integral of every continuous curve that passes through the points? Explain.

9. Is there a continuous curve that passes through the points and whose integral is negative? Explain.

In Part A, it appears that anything is possible.

Part B

In this part of the activity, let the vertical axis measure velocity with scale 1 unit = 10 mph and the horizontal axis measure time with scale 1 unit = 1 hr. Let all the graphs you drew in 2, 5, 6, and 7 represent the velocity of a car over a five-hour period.

1. Interpret the points given in 1 of Part A.

2. Interpret the curve you drew in 2 and the integral in 4 of Part A.

3. Interpret 6 of Part A.

4. Answer 8 of Part A in this context.

5. Answer 9 of Part A in this context.

In Part B, the realistic situation means that the possibilities are more restricted than they were in Part A. Expect to get answers like "a car can't travel faster than _____ miles per hour" or "no object can travel faster than the speed of light" as parts of answers to 4 of Part B. You should encourage and explain these answers in the context of a mathematical model.

Part C

The curve you found in 2 (Part A) has a derivative equal to 1 for all x. In this section assume all the curves you drew in Part A have a derivative at every point and the value of the derivative at each point is greater than or equal to 0.

1. Answer 8 of Part A.

2. Answer 9 of Part A.

In Part C, the mathematical restrictions imposed on the function mean that the possibilities are even more limited.

Numerical integration

Topics: Numerical integration, Simpson's rule

Summary: Shows how the parabolic rule can be "discovered" as a reasonable alternative to Riemann sums and the trapezoidal rule, using the special case of a quadratic function.

Background assumed: Riemann sums, trapezoidal rule.

Time required: 15 minutes.

Threads: approximation and estimation

1. Use six rectangles of equal base and the midpoint method to estimate the area in the first quadrant under the curve $y = 4 - (x-1)^2$ from $x = 0$ to $x = 3$.

Let some students or groups of students work on this problem using left endpoints, some using right endpoints and some using midpoints.

---

2. Let us define the error associated with an estimate to be the difference between the estimated value of the area and the exact value. That is

$$\text{error} = \text{estimate} - \text{exact value}$$

Compute the exact value of the area in question 1, and the error associated with the estimate that you computed.

Have the students compare the errors associated with the different estimates. Discuss the relationship between the error and the shape of the original graph.

---

3. Now estimate the same area, this time using six trapezoids. Again, compute the error associated with the new estimate.

Before the students begin to work, it is easy to have them develop the formula for the trapezoidal rule for this example. (They also should have no trouble generalizing it.)

---

4. Compare the errors associated with the midpoint and the trapezoidal methods. Devise a way to combine the two computations that will produce a better estimate than either of these.

They should observe that the absolute value of the error for the trapezoidal rule is twice that for the midpoint method, and that the signs are opposite. With minimal prodding, they suggest "two parts midpoint and one part trapezoidal" (a weighted average) as a superior estimate.

---

5. Compute the weighted average $\frac{(2\,\text{Mid} + \text{Trap})}{3}$. Compute the error associated with this estimate.

The error, of course, is zero. Point out that the weighted average always produces the exact value for a quadratic function; indeed, this method is equivalent to fitting parabolas to the three points on the graph that correspond to the endpoints and the midpoint of each subinterval.

Problems
Repeat this activity, this time estimating the area under the curve $y = x + \sin x$ from $x = 0$ to $x = \pi$, and using four subintervals. Choose the best estimate. Discuss how this problem differs from the original activity.

Simpson's rule is the most accurate, but it is not exact since $y = x + \sin x$ is not a parabola.

Verifying the parabolic rule

Topics: Numerical integration, Simpson's rule

Summary: Shows by example that Simpson's rule as developed in "Numerical integration" corresponds to the area under a parabolic segment.

Background assumed: Riemann sums, trapezoidal rule.

Time required: 15 minutes.

Threads: approximation and estimation

Let $f(x) = x - 1/x^2$. We will estimate the area under the curve from $x = 1$ to $x = 3$.

1. Draw a graph of the region.

2. Estimate the area using the midpoint method and one rectangle.

3. Estimate the area using one trapezoid.

4. Compute the weighted average $\dfrac{(2\,\text{Mid} + \text{Trap})}{3}$ (Simpson's rule, with $n = 2$).

5. The endpoints of the interval $[1,3]$ and the midpoint correspond to the three points $(1,0)$, $(2,7/4)$, and $(3,26/9)$ on the graph of $f(x)$. Verify that these three points also lie on the graph of the quadratic function
$$q(x) = \frac{11}{36}x^2 + \frac{30}{36}x - \frac{41}{36}.$$

6. Compute the area under the quadratic function in question 5, and compare your answer with the answer to 4.

7. On the same set of axes sketch the graphs of f and q between $x = 1$ and $x = 3$.

8. In your own words, describe what this activity illustrates.

The students should observe that the weighted average of rectangles and trapezoids is precisely the area under the parabolic segment. In other words, we are approximating the area under the curve by fitting a quadratic function and computing the area under it.

Finding the average rate of inflation

Topic: Integration

Summary: Students find the average value for a function presented as a graph.

Time required: 15 minutes

Thread: graphical calculus

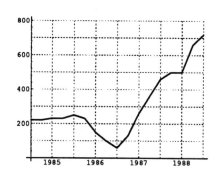

Inflation rate

Consider the graph, which gives the rate of inflation (in percent per year) in a country with economic difficulties for the years indicated.

1. Estimate the average rate of inflation over the time period shown.

One misconception that some students have with this problem is that they think the average rate is the slope of the line connecting the endpoints. (Since the word "rate" appeared in the problem some students thought that the problem involved finding the slope of the graph.) The instructor needs to emphasize that the concept involved in a problem is more important than a key phrase.

To steer the students in the right direction, either (1) look at a portion of the graph where the slope is negative and ask them if they think the average rate of inflation over that small section is negative. If they say yes, then ask them if prices are decreasing during that time period. Or (2) ask them, "If the graph represented the velocity of a car, how would you find the average velocity over the given time period?"

After they see that this is not a problem of finding the slope, point out that since the vertical axis measures a rate, the average will itself be a rate.

When the students worked on this there were two methods used:

1. The students took the difference between the rate at the end and the rate at the beginning and divided by 4. In geometric terms they found the slope of the line connecting the endpoints. Instead of simply saying this is incorrect, draw some graphs that have the same endpoints but obviously have different averages, then ask the students if they have the same average.

2. The students found the average by choosing a value in year 1 and then a value in year 2, etc., and then took the average of those values. This approach gives an approximate answer. Some students did it for 6-month periods. You can mention Riemann sums and then ask how to get a better approximation. The class should suggest using shorter time intervals.

You can finish up by relating this to area under a curve.

One good solution is to estimate the area by counting blocks—you will get a better estimate by using smaller blocks.

---

2. Draw a horizontal line with height equal to the average rate of inflation on the graph. Relate the area under the horizontal line to the area under the graph of the inflation rate.

Point out that the areas should be the same and that this is a graphical way to represent the average value of a function.

Problems

1. What was the average rate of inflation for the year 1986?

2. How much money would it take at the end of 1988 to have the same buying power as $100 at the beginning of 1985?

3. In a different country, the rate of inflation t years after 1986 is 4% times t.

    (a) What is the average rate of inflation for this country from 1985 to 1993?

    (b) Draw a graph of the rate of inflation versus time and indicate on the graph what the average is.

Inflation can be used to introduce the idea of exponential growth—ask what the average rate says about the cost of something in 4 years.

Cellular phones

Topics: Integration

Summary: Requires students to synthesize concepts of velocity and distance, and the integral as area under a curve, in a realistic situation.

Background assumed: Elementary integration

Time required: 25 minutes

Threads: distance and velocity, multiple representation of functions, graphical calculus

---

A car is traveling on a straight road on a stretch that contains cities A, B, C, and D as illustrated below. The car is between cities A and D.

```
   A          B          C          D
───┼──────────┼──────────┼──────────┼───
```

City A is 60 miles from City D, with City B 20 miles from cities A and C and City C 20 miles from city D. There are cellular phone receiving stations in each of the four cities. Each station has a range of ten miles.

1. Suppose that the car is traveling at a uniform rate of 55 miles per hour. What percentage of the time for the trip between cities A and D is spent within range of the station in City A? City B?

2. Suppose that the velocity of the car is

$$r(t) = \begin{cases} 60t, & t \leq 1 \\ 120 - 60t, & t \geq 1 \end{cases}$$

where t is measured in hours and $r(t)$ is measured in miles per hour. Also, suppose that at $t = 0$, the car is at City A. Now, what percentage of the time for the trip between cities A and D is spent within range of the station in City A? City B?

3. Do you think that it is possible to start driving from City A, stop driving at City D, and maintain the same percentage of time within range of each location? If so, draw a graph of the velocity of a car on such a trip. If not, explain why not.

Problems

1. Given the velocity of the car as described in part 2, do you think that you could move the location of the stations to other locations in such a way that the car would spend the same amount of time within range of each station? If so, justify your answer. If not, explain why not.

2. Below is a graph of a velocity of a car as it travels between City A and City D. What percentage of the time is spent within range of City B?

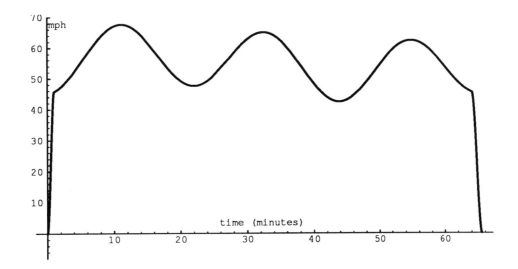

Velocity graph

This activity can be done with graphical calculus for a variety of velocity functions. Without graphical calculus, use of a computer algebra system might be required.

The shorter path

Topic: Arc length

Summary: This activity introduces the concept of arc length by using the distance formula. Discussion may lead students to realize the need for numerical integration.

Background assumed: Students need to know the formula for distance between two points and be familiar with integrals as limits of Riemann sums.

Time required: 20 minutes

Resources needed: None, although a computer or calculator that does numerical integration is very useful for the discussion following the activity.

Threads: approximation and estimation, modeling

The instructor may wish to do 2–5 of the assignment in class after finishing the activity or after assigning 1a and 1b.

1. Find the distance from the origin to the point whose coordinates are $(1,1)$.

2. Pat and Chris are hiking and they want to make camp at a campsite that is 1 mile east and 1 mile north from their current location. How far (as the crow flies) are they from the campsite? Write a brief explanation of how the answer to question 1 can be used to answer this problem.

The term "as the crow flies" is used to refer to the distance along a straight line between two points as opposed to the distance along the path between two points.

---

3. When Pat and Chris look at their map they see that there is a lake between the campsite and their current location. If they follow the path $y = \sqrt{x}$ (the Radical Road) they will skirt the northern edge of the lake and if they follow the path $y = x^3$ (the Cubic Camp Trail) they will skirt the southern edge of the lake.

    (a) Draw a picture of the situation.
    (b) Explain why both paths will reach the campsite.

4. Deciding to estimate the length of each path, they do the following. Each path has a rest stop located at a point on the path which is $\frac{1}{2}$ mile east and some distance north of the starting point.

    (a) How far north is each rest stop from their starting position?
    (b) How far (as the crow flies) is each rest stop from their starting point?
    (c) How far (as the crow flies) is each rest stop from the camp site?

    (d) Now give an estimate for the total distance along each path to the campsite. Is your estimate larger than the actual distance along the path (an upper estimate) or smaller than the actual distance along the path (a lower estimate) or neither? Explain your reasoning.

5. Is it possible for the hikers to hike less than 1.5 miles and get to the campsite using one of the paths? Write a brief explanation of your answer.

The estimates in 4d are about 1.445 and 1.523 miles respectively. The logic of using the estimates should be mentioned here. Since a lower estimate is 1.523 the cubical path must be longer than 1.5 miles. The answer to 4d does not *guarantee* that the radical route is less than 1.5 miles. You would need to find error estimates for the integral to be certain that the length was less than 1.5 miles. Students can find an upper estimate of 2 miles for the Radical Route by using the path from $(0,0)$ to $(0,1)$ to $(1,1)$.

6. Which path gives the shorter route to the campsite? Explain your choice.

Some students may confuse smaller area with shorter distances. For instance, in one of our classes some students found the area between the straight line connecting $(0,0)$ and $(1,1)$ and each of the two paths and then asserted that the path corresponding to the smaller area gave the shorter path. We drew a wiggly curve close to the straight line to show that a smaller area could correspond to a longer path. A nice research problem is to investigate under what conditions it is true that the smaller area guarantees a shorter path.

Problems

1. (a) If additional rest stops were added to each path at distances $\frac{1}{4}$ mile and $\frac{3}{4}$ mile east of the starting point, how far north are each of these rest stops from the starting point?
   (b) Use the additional rest stops to improve the estimates you found in 4d.

2. If rest stops are located every dx miles east of the starting point write an expression for an estimate of the distance along each path.

3. Use algebra or calculus to express your answer to 2 as a Riemann sum.

4. Express the distance along each path as an integral.

5. What is the distance along each path between the starting point and the campsite?

The integrals for the length of each path need to be evaluated numerically. The square-root path is approximately 1.47894 miles and the cubic path is about 1.54787 miles.

The River Sine

Topics: Integration, numerical integration

Summary: Students use numerical integration to estimate the cost of fencing a plot of land.

Background assumed: Numerical integration, arc length

Time required: 20 minutes

Resources needed: Calculator

Thread: approximation and estimation

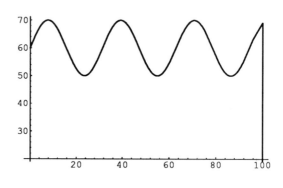

You own a plot of riverfront property which is pictured in the figure. Your property runs along the x-axis from $x = 0$ to $x = 100$ and is bounded by the lines $x = 0$, $x = 100$, and the River Sine whose equation is $y = 60 + 10\sin(x/5)$.

Your property

1. What is the area of the plot?

2. You have $600 to spend, and you hire a gardening firm to install a fence along the riverside part of your property. The firm will charge you $3 a foot for the fence including installation. If there is any money left over, the firm will fertilize the lot. They charge $20 a bag for fertilizer. (This charge includes spreading the fertilizer.) One bag of fertilizer will cover 1,000 square feet.

1. What is the area of the plot?

2. You have $600 to spend, and you hire a gardening firm to install a fence along the riverside part of your property. The firm will charge you $3 a foot for the fence including installation. If there is any money left over, the firm will fertilize the lot. They charge $20 a bag for fertilizer. (This charge includes spreading the fertilizer.) One bag of fertilizer will cover 1,000 square feet.

   (a) Will there be any money left for fertilizer after the fence along the river is installed? Explain your answer.

   (b) If there is money for fertilizer, about how much of your plot will be fertilized? Explain.

The problem does not explicitly say to find the length of the fence numerically, and many students do not use numerical techniques unless specifically told to do so in the problem. This is a good chance to emphasize that numerical integration should be used to evaluate integrals whenever an antiderivative can't be found.

Chapter 5

Transcendental Functions

Ferris wheel

Topics: Trigonometric functions and their derivatives

Background assumed: Some ideas about the graphical relationship between velocity and position.

Time required: 50 minutes

Threads: distance and velocity, graphical calculus, modeling

We will investigate the motion of a person riding on a Ferris wheel. The radius of the Ferris wheel is 40 feet. The Ferris wheel rotates in a counter-clockwise direction.

Part A

We will start with the simple case in which the wheel rotates through an angle of one radian each second.

---

1. At this time the instructor should review radian measure and mention that we will always use radian measure in calculus. If students ask why you can mention the fact that this unit for angles will make calculation much simpler. Be sure to pick up this thread when you get to the derivative of sine and cosine.

2. Remind the students that one of the main ways we investigate hard problems is to simplify the problem.

1. Set up a coordinate system with the origin at the center of the wheel. Let the point where the person is sitting be represented by the point whose coordinates are (40,0) to start.

Figure 1: Wheel

There will probably be some students who want the origin on the ground, but the class should believe that the center of the wheel is at least as good as that choice.

On your coordinate system, plot where the person will be 1 second later, $\pi/2$ seconds later, 2 seconds later, 3 seconds later, π seconds later, $3\pi/2$ seconds later, 2π seconds later.

What has happened in the first 2π seconds?

2. Now, on Figure 2, plot the horizontal position of the person for 10 seconds.

Be sure the students label units on the axes. Also, after they have plotted the 10 seconds ask them what would happen after that. They should all recognize that the graph repeats. The following exercise on the vertical position should be easy and clear up any confusion.

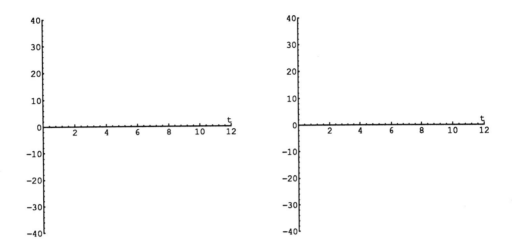

Figure 2: Horizontal   Figure 3: Vertical

3. On Figure 3, plot the vertical position of the person for 10 seconds.

4. Now go back and draw the horizontal and vertical positions on the same axes you used above, but assuming the wheel rotates at 2 radians per second.

5. Draw the graph of the horizontal position for 12 seconds if the wheel rotates at 1 radian per second for 8 seconds and then gradually increases its speed over the next 4 seconds to 2 radians per second and then continues to rotate at 2 radians per second.

If students have trouble with this after you explain it, assign the same problem for the vertical position.

---

6. Finally draw a graph of the horizontal position for what you think is a realistic Ferris wheel and explain your graph.

If students have trouble with this after you explain it, assign the same problem for the vertical position.

Part B

Now we will relate this work to functions we have already seen.

1. Draw a picture of the Ferris wheel with the person somewhere in the first quadrant. (We will call this point P.) Drop a perpendicular line to the horizontal axis from the point where the person is sitting. The point where this line hits the horizontal axis we will call Q.

2. Look at the right triangle formed by the points P, Q, and the origin, O.

    Denote the angle formed by QOP as θ and remember that the length of OP is 40 because the radius of the Ferris wheel is 40 feet.

    (a) Using trigonometry write an expression for the length of OQ. _____
    (b) Using trigonometry write an expression for the length of PQ. _____

3. Suppose the wheel is rotating at 1 radian per second.

    (a) Find a rule for θ in terms of time. $\theta =$ _____
    (b) Write a rule for the horizontal position of the person in terms of time. $x =$ _____
    (c) Write a rule for the vertical position of the person in terms of time. $y =$ _____

4. Suppose the wheel is rotating at 2 radians per second.

    (a) Find a rule for θ in terms of time. $\theta =$ _____
    (b) Write a rule for the horizontal position of the person in terms of time. $x =$ _____
    (c) Write a rule for the vertical position of the person in terms of time. $y =$ _____

The students should compare their rules with the graphs previously drawn.

If you wish you may introduce frequency and amplitude at this point. (Using a different size wheel should clarify the concept of amplitude and how it differs from frequency.)

---

5. (a) Draw a graph of the horizontal *velocity* of the person if the wheel is rotating at 1 radian per second.

Be sure the students recognize that the graph of the horizontal velocity is identical to the graph of the vertical position; i.e., the derivative of sine is cosine.

Also, find the times when the horizontal velocity is maximized, minimized, and equal to zero, and relate these to the graph of vertical position.

(b) Draw a graph of the vertical velocity of the person if the wheel is rotating at 1 radian per second.

Be sure the students recognize that the graph of the vertical velocity is the reflection of the graph of the horizontal position in the horizontal axis; i.e., the derivative of cosine is −sine.

Also, find the times when the vertical velocity is maximized, minimized, and equal to zero, and relate these to the graph of horizontal position.

Finally, look at the case where the wheel rotates at 2 radians per second and show that the horizontal velocity must be faster than in the 1 radian per second case as a preview of the chain rule.

Why mathematicians use e^x

Topic: Derivatives of exponential functions

Summary: Motivate the choice of e; using a calculator to estimate limits.

Background assumed: Definition of slope, definition of derivative, secant line; concavity is needed for the last part.

CHAPTER 5. TRANSCENDENTAL FUNCTIONS

Time required: 25–30 minutes (shorter version 15–20 minutes)

Resources needed: A calculator with a power (a^x) key

Thread: approximation and estimation

After the activity, the class should see that the derivative of a^x is a^x times a constant that depends on a. The number e is used because the corresponding constant is 1. A main point of this activity is that the students see that the choice of the base e for exponential functions makes the derivative of this exponential equal to the function without involving a messy constant factor (although e may be messy). You might want to use the analogy that using e is like using the metric system of measurement since it will make calculations easier.

You can turn this into a shorter activity by showing the students 1 and 7 and omitting 8.

---

1. (a) Apply the definition of derivative to write an expression for the derivative of the function $f(x) = 2^x$ at the point whose first coordinate is x.

   (b) Using the properties of exponents, simplify your answer to 1a, and show that it can be expressed as
   $$2^x \lim_{h \to 0} \frac{2^h - 1}{h}.$$

Notice that, if we can find the limit mentioned in 1b, we will know the derivative of 2^x at every point. But the limit in 1b is the slope of the tangent line to the graph of $y = 2^x$ at the point where $x = 0$.

2. Draw the graph of $y = 2^x$ with $x = 0$ in the center of your graph.

3. Draw the tangent line to the graph at the point $(0, 1)$ on the graph you drew in 2.

4. Now we will use the definition of derivative to find the slope of the tangent. On the graph you drew in 2 draw the secant line from $(0, 1)$ to the point on the graph of $y = 2^x$ whose first coordinate is 1.

   (a) Compute the slope of this secant line.
   (b) Is the slope of the secant line larger or smaller than the slope of the tangent line?
   (c) What do the results of 4a and 4b tell you about the slope of the tangent line to the graph at $(0, 1)$?

5. Next, on the graph you drew in 2, draw the secant line from $(0,1)$ to the point on the graph of $y = 2^x$ whose first coordinate is -1.

   (a) Compute the slope of this secant line.

   (b) Is the slope of the secant line larger or smaller than the slope of the tangent line?

   (c) What do the results of 5a and 5b tell you about the slope of the tangent line to the graph at $(0,1)$?

   (d) Using the results of 4 and 5, what can you now say about the slope of the tangent line to the graph at $(0,1)$?

6. Recall the definition of the derivative of the function $f(x) = 2^x$ at the point $x = 0$ is

$$f'(0) = \lim_{h \to 0} \frac{2^h - 1}{h}.$$

   (a) What value of h was used for the secant line you worked with in 4?
   (b) What value of h was used for the secant line you worked with in 5?
   (c) Rephrase your results of 5d in the language of derivatives.

7. Now we will fill in the tables below. (There is one table for positive values of h and a second table for negative values of h.)

positive h		negative h	
h	$\frac{f(0+h)-f(0)}{h}$	h	$\frac{f(0+h)-f(0)}{h}$
0.1		-0.1	
0.01		-0.01	
0.001		-0.001	
0.0001		-0.0001	

   (a) Fill in the table for $h = 0.1$ and $h = -0.1$. What can you now say about the derivative of $y = 2^x$ at $x = 0$?

   (b) Fill in the table for $h = 0.01$ and $h = -0.01$. What can you now say about the derivative of $y = 2^x$ at $x = 0$?

(c) Fill in the table for $h = 0.001$ and $h = -0.001$. What can you now say about the derivative of $y = 2^x$ at $x = 0$?

(d) Fill in the table for $h = 0.0001$ and $h = -0.0001$. What can you now say about the derivative of $y = 2^x$ at $x = 0$?

(e) How many decimal places do you now know of the value of $f'(0)$? Explain.

(f) What do you know about the derivative of 2^x?

---

1. Using $h = -1$ and $h = 1$ gives $1/2 < f'(0) < 1$.

   Using $h = -0.1$ and $h = 0.1$ gives $0.66967 < f'(0) < 0.71773$.

   Using $h = -0.01$ and $h = 0.01$ gives $0.69075 < f'(0) < 0.695555$.

   Using $h = -0.001$ and $h = 0.001$ gives $0.692907 < f'(0) < 0.69339$. Using $h = -0.0001$ and $h = 0.0001$ gives $0.69312 < f'(0) < 0.6932$.

2. The actual value of $f'(0)$ is $\ln 2$ ($= 0.693147...$). The slopes of the secant lines for negative h will form an increasing sequence that approaches $\ln 2$ as h approaches 0, while the slopes of the secant lines for positive h will form a decreasing sequence that approaches $\ln 2$ as h approaches 0.

3. For 7f, the students should see that the derivative of 2^x is a constant times 2^x and that the constant is $0.693....$

---

8. (a) What can you say about the slopes of the secant lines that correspond to negative h as h approaches 0?

(b) What can you say about the slopes of the secant lines that correspond to positive h as h approaches 0?

(c) Relate what you found in 8a and 8b to the concavity of the graph of $y = 2^x$.

(d) Summarize how you can estimate the value of the derivative at a point if you know the concavity of the graph in an open interval that contains that point.

Be sure that the students understand that negative values of h will give lower estimates for the derivative and positive values of h will give upper estimates for the derivative. To check that they understand this, draw a graph that is concave down and ask what type of estimate they will get for negative h.

Problems

1. Repeat parts 1, 2, 3, 5, 6, and 8 for the function $f(x) = 3^x$.

2. Repeat parts 1, 2, 3, 5, 6, and 8 for the function $f(x) = 4^x$.

3. If there is a number a with the property that the derivative of a^x at $x = 0$ is 1 what can you say about the number a? (Based on the results of the activity and problem 1 and problem 2.)

4. Decide whether the number a described in problem 3 is larger than 2.5. Explain your reasoning.

5. Decide whether the number a described in problem 3 is larger than 2.75. Explain your reasoning.

For problem 1, $f'(0) = \ln 3 \, (= 1.098612\ldots)$ and for problem 2, $f'(0) = \ln 4 \, (= 1.386294\ldots)$. For problem 4, $f'(0) = \ln 2.5 \, (= 0.91629\ldots)$ and for problem 5, $f'(0) = \ln 2.75 \, (= 1.01160\ldots)$.

Be sure to mention that the number a described in problem 3 is e. The problems use the idea of bisection to approximate the value of e. After completing 1 and 2, students will believe that $2 < e < 3$; after completing 4, they know that $2.5 < e < 3$; and after completing 5, they will know that $2.5 < e < 2.75$.

Exponential differences

Topic: Derivatives of exponential functions

Summary: By using finite differences, the students discover that the derivative of an exponential function is a constant times the exponential function.

Time required: 10 minutes

Threads: approximation and estimation, multiple representation of functions

See also "Finite differences."

---

1. Each member of the group should choose a different exponential function with integer bases (e.g., $y = 3^x$, $y = 4^x$, etc.) and construct a finite difference table as is illustrated in the example below. That is, if your function is $y = f(x)$, in Row 1 compute $f(1), f(2), f(3), f(4), f(5), f(6)$. In Row 2, over the star "*" compute $f(2) - f(1)), (f(3) - f(2)), (f(4) - f(3)), (f(5) - f(4))$, and $(f(6) - f(5))$. In Row 3 and in subsequent rows, over the stars, compute the difference between the entry in the row directly above the star to the right and the entry in the row directly above the star to the left.

1	2	3	4	5	6	Row 0
*	*	*	*	*	*	Row 1
	*	*	*	*	*	Row 2
		*	*	*	*	Row 3
			*	*	*	Row 4
				*	*	Row 5
					*	Row 6

Example: $f(x) = 2^x$

1	2	3	4	5	6	Row 0
2	4	8	16	32	64	Row 1
	2	4	8	16	32	Row 2
		2	4	8	16	Row 3
			2	4	8	Row 4
				2	4	Row 5
					2	Row 6

As a group, determine any similarities in the patterns that develop. Can you describe what happens in general? (That is, what happens with a general exponential function $y = a^x$?)

Discuss this before going on.

2. If we combine the process of taking differences with division by entries in the previous row, the rows below row 1 become constant. By what entries in the previous row should you divide? How does the constant relate to the base of your exponential function?

Notice that the rate of change of the function appears to be proportional to the value of the function. In the tables, Row 2 is obtained from Row 1 by computing $\dfrac{(a^{n+1} - a^n)}{1}$ or $\dfrac{(a^{n+h} - a^n)}{h}$ where $h = 1$. Thus, this number is an approximation to $f'(n)$.

Follow up:

Have the students simplify $\dfrac{(a^{n+h} - a^n)}{h}$ to obtain $a^n \dfrac{(a^h - 1)}{h}$. Then have them use calculators to approximate $\lim_{h \to 0} \dfrac{(a^h - 1)}{h}$ for various numbers a. Observe that the values approach $\ln a$.

See also "Why mathematicians use e^x."

Inverse functions and derivatives

Topics: Inverse functions, derivatives of inverse functions

Summary: Uses two examples of realistic functions with inverses and relates the slope of the inverses to the slope of the function

Time required: 30 minutes

Threads: modeling, graphical calculus

The inverse of a function f, if it exists, is another function, g, that reverses the action of f. For example, if the function f gives the position, s, of an object as a function of time, t, then the inverse function, g, gives the time as a function of the position. That is, given any position, s, the function g gives the time, t, when the object is at position s.

In this activity, we will investigate inverse functions and their derivatives.

1. The function, f, whose graph is below gives the number of German marks one could obtain for x U.S. dollars on October 1, 1993.

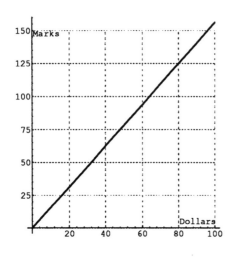

Dollars to Marks

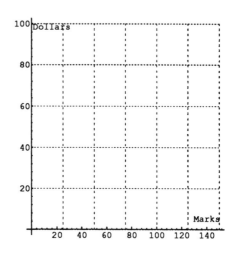

Marks to Dollars

(a) If I held $20 US on October 1, how many German marks could I obtain?

(b) If I held U.S. dollars on October 1 and wanted to obtain German marks, what exchange rate was I dealing with? Explain how the exchange rate can be read from the original graph.

To get more accurate answers, use points with both coordinates on the lattice (if possible). For example, use $(0,0)$ and $(80, 125)$ to get the exchange rate of $\frac{125}{80} = 1.5625$ marks per dollar.

(c) If I held 40 German marks on October 1, how many U.S. dollars could I obtain?

It appears that the answer is about 60–65 marks. Have them check by using the exchange rate: $40(1.5625) = 62.50$ marks.

(d) On the second set of axes above, sketch the graph of the function, g, that gives the number of U.S. dollars one can obtain for x German marks. That is, sketch the graph of the function, g, that is the inverse of the given function, f.

(e) If I held German marks on October 1 and wanted to obtain U.S. dollars, what exchange rate was I dealing with? Explain how the exchange rate can be read from the graph you just sketched.

(f) How are the two exchange rates you found related?

(g) Describe the relationship between the two exchange rates using the language and notation of calculus.

Be sure to observe that the functions involved in this part are linear, so the derivatives do not depend on the point at which they are evaluated. When you continue with the next part, point out how the non-constant derivative requires a more complete description of the relationship.

---

2. An athlete, Jane, running a five-mile course has her coach plot her position as a function of time. The position is the distance, in miles, that she has run from the start of the course, and time is measured in minutes from the time she started her run. Call this function f. The graph is shown below.

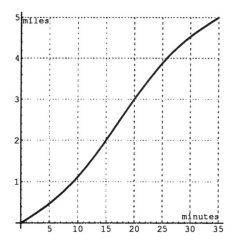

Miles vs. Minutes

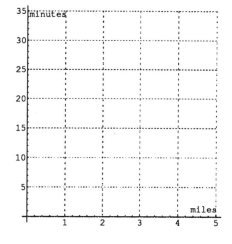

Minutes vs. Miles

(a) How long did it take Jane to run the first mile? the first two miles? the first three miles? all five miles?

(b) How fast, in miles per minute, was she running after ten minutes? Explain how to read this value from the original graph. How can this speed be described precisely, using the language and notation of calculus?

Her velocity appears to be about $2/15 = .133$ miles per minute, which is the value of $f'(10)$.

(c) Most runners, for practical purposes, measure their speed in minutes per mile and think of a course in terms of miles traveled so far and how long it took them to reach each milepost. On the second set of axes above, make a plot of Jane's time as a function of the distance traveled so far. Call this function g. It is the inverse of the original function, f.

(d) How fast, in minutes per mile, was Jane running at the end of the second mile? Explain how to read this value from the graph you just drew. How can this speed be described precisely, using the language and notation of calculus?

Her velocity appears to be about $15/2 = 7.5$ minutes per mile, which is the value of $g'(2)$.

(e) Describe carefully, using the language and notation of calculus, how the two rates you found above are related.

We have $g'(2) = \dfrac{1}{f'(10)}$, and since $10 = g(2)$, this can be written $g'(2) = \dfrac{1}{f'(g(2))}$. Be sure the students see that $g'(2)$ is *not* $\dfrac{1}{f'(2)}$.

Fitting exponential curves

Topic: Exponential functions

Summary: Students use the semi-log scale to graph the exponential function.

Background assumed: Students need to know properties of ln.

Time required: 15 minutes

Resources needed: A calculator with a ln button.

Threads: approximation and estimation

Consider the following data.

x	y
0	3
1	3.66
2	4.48
3	5.47
5	8.15
6	9.96
8	14.86
10	22.17
12	33.07
15	60.27

1. Sketch these points on a set of axes, and decide whether the function is increasing at a constant rate, increasing and concave up, or increasing and concave down. Explain you choice.

2. Calculate the natural logarithm of the y values. Now on a new set of axes plot $\ln y$ versus x. What do you notice?

3. If a function has the form $y = ce^{bx}$, we say it is an exponential function. Take the natural logarithm of both sides of the above equation to show that $\ln y = \ln c + bx$.

4. If we let $Y = \ln y$ and $a = \ln c$ in the above we have that $Y = a + bx$, which says that Y is a linear function of x. That is, if we plot Y versus x we should get a straight line. Use your ruler to draw a straight line through the points in your graph from part 2.

5. Estimate the slope and the intercept for the straight line you drew in part 4.

6. Using your answers from part 5 you now have estimates for a and b. This will also allow you to find c since $a = \ln c$. (Do you see why?) Now use these values to find an exponential function that fits the original data.

To summarize, we note that if we think a table of data might have come from an exponential function, we can plot the logarithms of the y values versus the x values and see if it looks like a straight line. We also note that the straight line we get also allows us to estimate the parameters c and b.

Problems

The following gives the population of Mexico for the years 1980 to 1986.

Year	Population (in millions)
1980	67.38
1981	69.13
1982	70.93
1983	72.77
1984	74.66
1985	76.60
1986	78.59

It has been claimed that the population of Mexico grew exponentially during this period. What do you think? If you decide the answer is yes, find the exponential function which fits this data. What was the growth rate for the population during this time period? What is your estimate for the population in 1990? 1995?

This activity involves an important concept in curve fitting. Namely, if two variables x and y are related in an exponential way and if we plot $\ln y$ versus x, then we get a straight line. In particular, let students spend time on parts 1 and 2 before working on subsequent parts. This concept also arises when you study the basic Malthus model of population growth and its related differential equation. When you solve $dy/dx = ky$ by separation of variables, you reach a point where you have $\ln y = kx + c$, which implies that y and x are related in an exponential way.

Log-log plots

Topics: Logarithmic, exponential, and power functions; scaling and curve fitting

Summary: Uses change of scale to determine whether variables are related by a power function.

Background assumed: Logarithmic and exponential functions

Time required: 30 minutes

Resources needed: Calculator with a ln button

Threads: multiple representation of functions, approximation and estimation

---

Consider the following table of data.

x	y
1	3
2	8.49
3	15.59
5	33.54
6	44.09
8	67.88
10	94.87
12	124.70
15	174.28

1. Sketch these points on a set of axes and decide whether the function is increasing at a constant rate, increasing and concave up, or increasing and concave down. Explain your choice.

2. Calculate the natural logarithm of the x and y values. On a new set of axes plot $\ln y$ versus $\ln x$. What do you observe?

3. If a function has the form $y = cx^b$, $b > 0$, we say it is a power function. Take the natural logarithm of both sides of the above equation to show that $\ln y = \ln c + b \ln x$.

4. If we let $Y = \ln y$, $A = \ln c$, and $X = \ln x$, we have that $Y = A + bX$, which says that Y is a linear function of X with slope = _____ and intercept = _____. That is, if we plot Y versus X we should get a straight line. Use your ruler to draw a straight line through the points in your graph from part 2.

5. Estimate the slope and the intercept for the straight line you drew in part 4.

6. Using your answers from part 5 you now have estimates for A and b. Use these estimates to find c. Now you have an approximation of the power function which fits the original data.

7. Use your approximation from part 6 to calculate approximate y-values for all the x-values in the table of data. Compare the computed y-values with the given y-values. How good is your approximation?

8. Write a brief summary of how we might test whether data may have come from a power function. Also indicate how you might estimate the power function.

Problem

Fit a power curve to the following data:

x	y
2	1.13
3	3.12
5	11.18
7	25.93
8	36.20
10	63.25
12	99.77
14	146.67

The log-log scale is used to graph data from an unknown power function. The unknown function is estimated and compared to the original data. This activity allows the instructor to introduce change of scale and curve fitting into the course at an early stage. This activity and "Fitting exponential curves" make a nice pair.

Using scales

Topics: Semi-log and log-log plots; scaling and curve fitting

Summary: Uses semi-log and log-log scales to identify exponential and power functions.

Background assumed: It helps to have done "Fitting exponential curves" and "Log-log plots" before doing this activity.

Time required: 20 minutes

Resources needed: Calculator with e^x and a^x buttons

Threads: multiple representation of functions, approximation and estimation

---

Beneath are three sets of coordinate axes. The first is a normal cartesian or x–y coordinate system. The second set is called *semi-log* since the vertical scale is logarithmic. (So in normal coordinates you would be plotting points whose form was $(x, \ln y)$ instead of (x, y).) The third is called *log-log* since both the vertical and horizontal scales are logarithmic. (This would be like plotting points of the form $(\ln x, \ln y)$ instead of (x, y).)

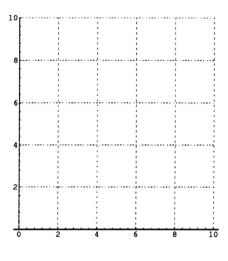

Cartesian

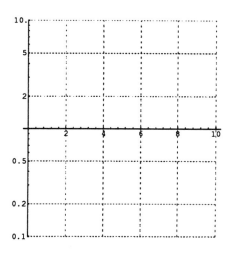

Semi-log

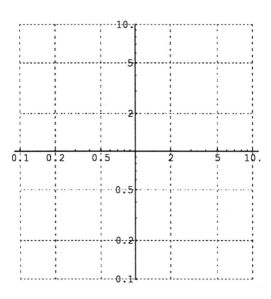

Log-log

Plot each of the relations on *all three axes*.

$$y = x + 2$$
$$y = 2e^{.2x}$$
$$y = 3x^{.5}$$

1. There should be at least one set of axes where each relation looks linear. Identify each relation and the scale that makes it linear. Explain why that scale makes the graph linear.

    The choice of points will make a difference, since $\ln t$ is undefined for $t \leq 0$.

    Since $\ln(x+2)$ is very close to $\ln(x)$ as x tends to ∞, the relation $y = x + 2$, which is linear in normal coordinates, will also appear very close to linear on the log-log scale.

    ---

2. Summarize your findings in a brief paragraph.

Problems

1. You expect some data to follow the model $y = Cx^\alpha$.

    (a) What scale should you use to plot the data?

The data should be linear on the log-log scale.

(b) Explain how would you estimate α.

We can estimate α by calculating $\dfrac{\ln y_1 - \ln y_2}{\ln x_1 - \ln x_2}$ where (x_1, y_1) and (x_2, y_2) are two points on the curve—that is, by finding the slope of the graph of $\ln y$ as a function of $\ln x$.

(c) Explain how you would estimate C.

Since $\ln y = \ln C + \alpha \ln x$, you can find $\ln C$ by using your estimate for α and any point (x, y) on the curve. Once you have an estimate for $\ln C$, use the exponential function to get C.

2. (a) Look at the census data for the US from 1790 to 1860. Find the best model to fit the data.
   (b) Use your model to predict the population in 1870 and 1990.
   (c) Check with the actual population of the US in 1870 and 1990. Explain any discrepancies.

Note: "Fitting exponential curves" gives more details on using semi-log plots, and "Log-log plots" gives more details on using log-log plots.

Chapter 6

Differential Equations

Direction fields

Topic: The geometric relationship between a differential equation and its solution.

Summary: Students develop the geometric method of using direction fields or slope fields to approximate the solutions to a differential equation. The notion of isoclines is used in this development.

Background assumed: Introduction to differential equations—students should realize that a first-order differential equation specifies the slope of a solution at each point.

Time required: 40 minutes

Threads: graphical calculus, approximation and estimation

This activity details a geometric method of solution for first-order differential equations. This method is called the method of direction fields and is based on the fact that a differential equation (DE) gives a direction at each point in the plane. For example, the DE $\frac{dy}{dx} = x$ says the slope of the curve $y(x)$ at the point (x, y) in the x–y plane equals the x coordinate. So the slope of y at the point $(1, 0)$ equals 1. In fact, the slope at any point $(1, y)$ equals 1.

We can use this information to approximate the solution curves to the given DE. We see from the above discussion that along the vertical line $x = 1$ the slope has the value 1. This line is called an *isocline*. For this differential equation, all the isoclines (curves of same slope) are vertical lines.

1. For this differential equation, find the isocline where the slope is -1.

2. For this differential equation, find the isocline where the slope is 0.5.

Suppose now that $\frac{dy}{dx} = x + y$.

3. Show that the isoclines of this DE are also straight lines. Find the lines where the slope is 1, −1, and 0.5.

Consider the original DE $\frac{dy}{dx} = x$.

4. On an axis draw the vertical line $x = 1$ and put a small dashed line of slope 1 at a few points on that to line indicate that the solution curve has slope 1 for all points on this line. What kind of dashed lines will you place on the vertical line $x = -1$? How about $x = 0.5$?

5. Continue in this fashion until you can estimate what the solution curves look like. What kinds of assumptions are you making here?

6. Find the solution that passes through the point $(0, 1)$.

Problems

1. Use this method to estimate the shape of the solution to the DE $\frac{dy}{dx} = x + y$ that passes through the point $(0, 1)$. What happens to this solution as x gets big?

2. Let $\frac{dy}{dx} = \frac{y}{x^2}$, $x \neq 0$. What are the isoclines for this DE? Sketch a solution curve that passes through the point $(1, 1)$.

Using direction fields

Topics: Differential equations, direction fields

Summary: Students draw a direction field and use it to draw and analyze solutions to the logistic equation. This activity illustrates how the use of a direction field enables us to get a good idea of the solution of a differential equation without doing lots of algebra.

Background assumed: Introduction to direction fields

Time required: 20–25 minutes

Threads: graphical calculus, approximation and estimation

Consider the differential equation described by

$$\frac{dy}{dt} = y(100 - y).$$

1. Find any constant solutions to the differential equation.

2. Draw some direction vectors for the differential equation.

3. Sketch the solution to the differential equation that passes through $y = 10$ when $t = 0$.

4. Write a short paragraph that describes the solution you found in part 3.

5. What will happen to the solution you found in part 3 as $t \to \infty$?

Problems

1. Sketch the solution to the differential equation that passes through $y = 150$ when $t = 0$.

2. Write a short paragraph that describes the solution you found in 1.

3. What will happen to the solution you found in 1 as $t \to \infty$?

4. Try to classify *all* solutions to the equation by their behavior as $t \to \infty$.

---

1. Emphasize that the scale for the graph should be chosen *after* the constant solutions are found.

2. A few useful properties of the solutions worth pointing out are:

    (a) The solution $y = 100$ attracts solutions, and

    (b) The solution $y = 0$ repels solutions.

3. This is an example of a logistic equation, which is one of the basic models in population biology. Interpret the differential equation in terms of population (the rate of change of the population is equal to the population times the difference between 100 and the population), and explain that as the population gets near 100 (the *carrying capacity*) the rate of change of the population approaches zero, which means that the solution curves will flatten out.

Drawing solution curves

Topic: The geometric relationship between a differential equation and its solution.

Summary: Given a number of direction fields, students are asked to plot approximate solution curves.

Background assumed: Significance of direction field

Time required: 15 minutes

Threads: approximation and estimation, graphical calculus

For each of the direction fields given below:

1. Draw a solution curve through the point marked on the direction field.

2. Draw a solution that passes through $y = 1$ when $t = 0$.

3. Draw a solution that passes through $y = 3$ when $t = 2$.

4. List any constant solutions.

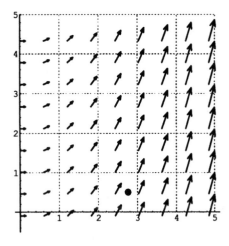
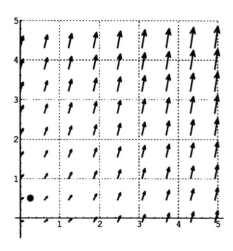

A   B

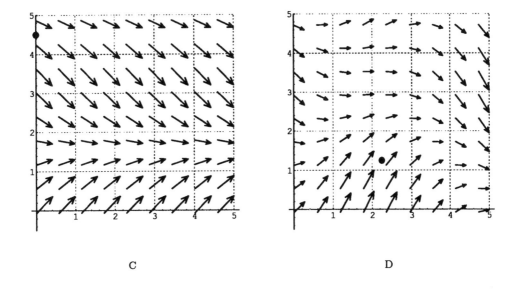

C D

The functions are $\frac{dy}{dt} = t$, $\frac{dy}{dt} = y + t$, $\frac{dy}{dt} = \cos y$, $\frac{dy}{dt} = \sin t + \cos y$.

The hot potato

Topics: Linearity, derivatives

Summary: This introduces Euler's method of approximating solutions of differential equations. Using the fact that a first-order differential equation gives the slope of the solution curve at each point, we approximate the solution.

Background assumed: Derivative as slope, introduction to differential equations and their solutions

Time required: 30 minutes (longer if you plan to talk about Newton's law of cooling)

Threads: approximation and estimation, modeling, graphical calculus

Newton's law of cooling states that a body cools at a rate directly proportional to the difference between the body's temperature and the temperature of the surrounding medium.

This law can be written as
$$\frac{dT}{dt} = k(T - A)$$
where $T(t)$ is the temperature of the body at time t, A is the temperature of the surrounding medium (the *ambient temperature*, which we assume to be constant), and k is the constant of proportionality.

A hot potato of 300° F is placed on the kitchen table in a room kept at 72° F. The potato begins to cool according to Newton's law of cooling. Let us assume that $k = -0.12$, so the following differential equation governs this situation:
$$\frac{dT}{dt} = -0.12(T - 72) \qquad (*)$$
where $T(t)$ is the temperature of the potato at time t, and in particular $T(0) = 300°$ F, and time is measured in minutes.

1. If we substitute $t = 0$ into (*) we get $\frac{dT}{dt} = -0.12(T(0) - 72) = -0.12(300 - 72) = -27.36$ degrees per minute. What does this tell us about how fast the potato is cooling when it is first put on the kitchen table?

2. Estimate the temperature of the potato after it has been on the table for one minute.

3. Estimate the rate of change of temperature after one minute.

4. Use the information from parts 2 and 3 to estimate the temperature of the potato after it has been on the table for two minutes.

5. Estimate the temperature of the potato after it has been on the table for five minutes.

6. Try to find a step-by-step method to find the temperature of the potato at any time knowing its temperature at a particular time and knowing that it satisfies (*) all the time.

Problems

1. How accurate do you think your method in part 6 is? What could you do to improve the accuracy of your method?

2. Check that $T(t) = 72 + 228e^{-.12t}$ is a solution to (*) satisfying $T(0) = 300°$. Use this rule to check the accuracy of your answers to the above parts.

3. Sketch a graph of temperature T versus time t. What does the concavity of this graph say about how an object cools if the object obeys Newton's law of cooling? Physically does this make sense?

This activity introduces Euler's method, a basic technique for approximating solutions of differential equations. Some time could be spent on the modeling of the differential equation, but the bulk of the time should be spent on the qualitative information that a differential equation gives about its solution. In particular, slope and linearity are used to construct the solution in a step-by-step fashion.

Spread of a rumor: discrete logistic growth

Topics: Differential equations, numerical solutions

Summary: Students solve a discrete logistic problem.

Time required: 30 minutes

Resources needed: Calculator

Threads: multiple representation of functions, approximation and estimation, modeling

---

A rumor spreads at a rate that is proportional to the product of the number of people in a population who have already heard the rumor (hearers) and the number of people who have not heard the rumor (ignorants). Using a discrete model for the spread of the rumor involves choosing a unit of time, Δt, and estimating how many additional people hear the rumor (converts) in each time period.

1. Using a discrete model for the spread of the rumor, fill in the table on the next page. (The numbers in parentheses assume rounding to the nearest integer at every step.)

The students should be concerned with whether or not to round the values at all the intermediate steps. Have some students round and others maintain two or three decimal places, and compare the differences that result. If students are working in teams, have members of each team use different rounding methods.

Day	# hearers(y)	# ignorants ($1000 - y$)	# converts (Δy)
0	5	995	4.975 (5)
1	9.975 (10)	990.025 (990)	9.875 (10)
2			
3			
4			
5			
6			
7			
8			
9			
10			
11			
12			
13			
14			

$$\begin{aligned}
\text{Population} &= 1000 \\
y &= \text{Number of hearers} \\
1000 - y &= \text{Number of ignorants} \\
\text{Model: } \frac{\Delta y}{\Delta t} &= 0.001 y (1000 - y) \\
\text{or } \Delta y &= 0.001 y (1000 - y) \Delta t \\
\text{Step size, } \Delta t &= 1 \text{ day} \\
\text{so, } \Delta y &= 0.001 y (1000 - y)
\end{aligned}$$

2. Plot the number of hearers, y, as a function of the number of days since the start of the rumor.

The graph should look like a discrete logistic curve.

---

3. Explain in words what the graph and the table indicate about how the spread of the rumor takes place.

Follow up this activity with a discussion of the analogous continuous model: $y' = ky(P - y)$. A lot can be learned about this model by examining properties of y' and y'', without solving the differential equation. If the students have learned to integrate using partial fractions, have them solve the differential equation. If not, give them the solution and have them verify that it is a solution. In either case, have them compare the number of hearers on day 8 for the analytic solution and the simulated solution. Also, have them compare the long-run behavior of the solutions.

Population

Topics: Differential equations, exponential growth

Summary: Students solve a differential equation and analyze the basic exponential growth model. The second question adds emmigration to the model.

Background assumed: Separation of variables

Time required: 25 minutes

Thread: modeling

---

1. The birth rate in a state is 2% per year and the death rate is 1.3% per year. The population of the state is now 8,000,000.

    (a) At what rate are babies being born in the state now? Be sure to include units of measure in your answer.

    (b) At what rate are people dying in the state now?

    (c) Write a differential equation that the population of the state satisfies. Be sure to define your terms.
    (d) Solve the differential equation you wrote in part 1c.
    (e) In how many years will the population reach 10,000,000?
    (f) Does the population have a steady state? Explain.

2. We will now try a more realistic model for the population. The birth rate and death rate are the same as in 1 and the population now is 8,000,000; however people are moving out of the state at a constant rate of 50,000 people per year.

    (a) Write a differential equation for the population and solve it.

(b) Will the population ever reach 10,000,000? If so, when? If not, why not?

Problems

1. For the situation described in 2:
   (a) Will the population ever reach 6,000,000? If so, when? If not, why not?
   (b) Does the population have a steady state? Explain.
   (c) At what constant rate will people have to leave the state in order for the state to have a constant population?

Some students will answer question 2 incorrectly by taking the solution to question 1 and subtracting $50,000t$. A concrete illustration of the problem with this approach is to point out that once people leave the state, then their children will not be born in the state.

Save the perch

Topic: Differential equations

Summary: Students model an elementary input-output situation with a differential equation.

Background assumed: Students need to be able to solve simple first-order differential equations. (Separation of variables is sufficient.)

Time required: 25 minutes

Thread: modeling

Happy Valley Pond is currently populated by yellow perch. The pond is fed by two springs: spring A contributes 50 gallons of water per hour during the dry season and 80 gallons of water per hour during the rainy season. Spring B contributes 60 gallons of water per hour during the dry season and 75 gallons of water per hour during the rainy season. During the dry season an average of 110 gallons of water per hour evaporate from the pond and an average of 90 gallons per hour of water evaporate during the rainy season. There is a small spillover dam at one end of the pond and any overflow will go over the dam into Bubbling Brook. When the pond is full (i.e., the water level is the same height as the top of the spillover dam) it contains 475,000 gallons of water.

Spring B has become contaminated with salt and is now 10% salt. (This means that 10% of a gallon of water from Spring B is salt.) The yellow perch will start to die if the concentration of salt in the pond rises to 1%. Assume that the salt will not evaporate but will mix thoroughly with the water in the pond. There was no salt in the pond before the contamination of spring B. Your group has been called upon by The Happy Valley Bureau of Fisheries to try and save the perch.

Part A

Let $t = 0$ hours correspond to the time when Spring B became contaminated. Assume it is the dry season and that at time $t = 0$ the pond contains 400,000 gallons.

1. Let $P(t)$ be the amount of water in the pond at time t. What will the change in the amount of water in the pond be during the time interval t to $t + \Delta t$ for some (small) positive number Δt?

2. Using 1, find a differential equation for P.

3. Solve the differential equation and decide if the pond will fill up. If so how long after $t = 0$ until the pond is full?

Part B

Use the same assumptions as Part A.

1. Let $S(t)$ be the amount of salt in the pond at time t. What will the change in the amount of salt in the pond be during the time interval t to $t + \Delta t$ for some (small) positive number Δt?

2. Using 1, find a differential equation for S.

3. Solve the differential equation and decide what will happen to the amount of salt in the long run.

4. Draw a graph of the amount of salt in the pond versus time for the next three months.

5. How much salt will there be in the pond in the long run?

6. Do the fish die? If so when do they start to die?

Part C

Answer the questions in Part B if the pond was full at time $t = 0$ and it was the dry season.

Problem

Answer the questions in Part B if the pond was full at time $t = 0$ and the contamination of Spring B occurred during the rainy season.

This activity develops students' ability to model situations using differential equations. The equation for part A is $dP/dt = 0$, for parts B and C, the equation is $dS/dt = 6$. The assigned problem is a mixture problem and is more difficult; the equation is $dS/dt = 7.5 - (75/475000)S$.

Chapter 7

Series

Convergence

Topic: Asymptotic behavior

Summary: Students answer a few open-ended questions about possible behavior of functions at infinity.

Background assumed: Derivatives, concavity

Time required: 20 minutes

Thread: approximation and estimation

Decide whether or not each of the following statements is true *all the time*. If you think the statement is true write an explanation or a proof. If you decide it is not true give an example of a function that shows the statement is not true.

1. If the derivative of f is positive for all $x > 0$, then $f(x) \to \infty$ as $x \to \infty$.

2. If $f'(x) \to 0$ as $x \to \infty$ then $f(x)$ converges to some *finite* number as $x \to \infty$.

Comments:

1. This activity can be done either graphically or analytically. The instructor should show both types of answers.

2. This activity can help students overcome the confusion between the convergence of the nth term of a series and the convergence of the series. You can use the analogy that $\sum a_n$ corresponds to f and a_n corresponds to f'.

Investigating series

Topic: Series of constants

Summary: Introduces the idea of partial sums and convergence of sequences of partial sums

Background assumed: Students should be familiar with convergence of numerical sequences.

Time required: 20 minutes

Resources needed: If you are inclined to use technology, you need something that can perform rational arithmetic.

Thread: approximation and estimation

In this activity, you will experiment with some infinite sequences and their limits. Starting with a given sequence of numbers, $\{b_1, b_2, \ldots\}$, you will construct a new sequence $\{a_1, a_2, \ldots\}$ as follows:

$$\begin{aligned} a_1 &= b_1 \\ a_2 &= b_2 - b_1 \\ a_3 &= b_3 - b_2 \\ &\vdots \\ a_n &= b_n - b_{n-1} \\ &\vdots \end{aligned}$$

Starting with the following sequence as $\{b_n\}$:

$$\frac{2}{1}, \frac{8}{3}, \frac{26}{9}, \frac{80}{27}, \frac{242}{81}, \frac{728}{243}, \frac{2186}{729}, \ldots$$

1. Compute the first six elements of the sequence $\{a_n\}$.

2. Graph $\{a_n\}$ versus n and $\{b_n\}$ versus n on the same set of coordinate axes. Plot at least the first six values for each sequence. Visually determine the limit of each sequence, if it exists, and place it on the same graph as a horizontal asymptote.

3. Find an expression for b_n and one for a_n in terms of n.

4. Compute the limit of $\{b_n\}$ as $n \to \infty$ and the limit of $\{a_n\}$ as $n \to \infty$. Compare these with the limits you found in 2.

5. The definition above gives a_n in terms of b_n and b_{n-1}. Using this definition, write an expression for b_n in terms of just the a_i's.

6. Use your answers to 4 and 5 to explain in your own words how the sequence $\{a_n\}$ is related to the sequence $\{b_n\}$.

7. Explain in your own words how the limit of $\{b_n\}$ as $n \to \infty$ is related to the sequence $\{a_n\}$.

Problem
Repeat the activity, this time starting with the following sequence as $\{b_n\}$:

$$\frac{3}{4}, \frac{6}{6}, \frac{9}{8}, \frac{12}{10}, \frac{15}{12}, \frac{18}{14}, \frac{21}{16}, \ldots$$

Note that the series in this problem is not geometric.

Space station

Topic: Limit of a series

Summary: This is one way to introduce convergent series.

Background assumed: Limit; series ideas *not required*

Time required: 10–15 minutes

Thread: approximation and estimation

A space ship is heading towards a space station that is 2.5 miles away. If the space ship travels 1 mile in the next second and then $\frac{1}{2}$ as far each second as in the previous second will it hit the station? If so, when?

Problems

1. How long will it take the space ship to travel 1.9 miles? How long to travel 1.99 miles? How long will it take until the space ship is .51 miles from the space station?

2. If the space ship travels 3/4 as far in each second as the previous second would it hit the space station? If so, when?

3. If the space ship travels x times as far in each second as the previous second, give an expression involving x that tells how far the space ship travels in five seconds. How far does it travel in n seconds? What values of x would result in the ship hitting the space station? If so, when?

4. Summarize these results in terms of the geometric series
$$1 + x + x^2 + x^3 + x^4 + \cdots.$$

Comments:

1. The class will usually split into two groups: those who say yes because the ship will always travel a positive distance forever and those who say no because it will only travel a finite distance. Arguments between these two groups will recap a lot of the history of mathematics. You may want to have Zeno's paradoxes available as a reference.

2. The following is a guided discovery approach. First have the students make three columns: time, distance traveled in that second, and total distance. Have the students fill in the first three or four rows of each column. The third column will be a partial sum. You can point out that it is easy to get the total distance after the next second by translating the previous entry into one with the denominator of the current row and adding 1 to the numerator. (E.g., after three seconds the space ship will have traveled $1 + 1/2 + 1/4 = 7/4$ miles; in the fourth second it will travel $1/8$ of a mile so write $7/4$ as $14/8$ and the total distance is $14/8 + 1/8 = 15/8$ miles. After doing a few more lines of the table, ask the class what the total distance is after n seconds. The class should easily see the pattern that the total distance after n seconds is $2 - \frac{1}{2}^{n-1}$ miles. Once they have this formula it is easy to see what happens as $n \to \infty$, and they will have a much clearer idea of convergence.

3. We teach differential equations before series, so some students tried to treat this problem as a differential equation. This is a nice way to relate the concepts of differential equations and series. If you assume that the velocity after t seconds is 2^{-t} miles per second, then the series is like using a Riemann sum for an integral.

Decimal of fortune

Topics: Infinite series, convergence

Summary: This activity provides an introduction to partial sums of series with positive and negative terms. It can serve as a springboard to standard results about alternating series.

Background assumed: It is suggested that this activity be done during a discussion of infinite series, after the divergence of the harmonic series has been discussed, but before alternating series have been discussed.

Time required: Each game takes about five minutes. Time for answering the questions should be five to ten minutes.

Resources needed: Calculator

Thread: approximation and estimation

What follows is a description of a game for two people, Player A and Player B. The object of the game is for Player B to determine a number that has been selected by Player A (and is unknown to Player B as the game begins). A score is computed based on how close Player B has come to Player A's number at the end of the game. Play requires a calculator.

Player A writes down a decimal with eight decimal places. This decimal must be between zero and one. Player B enters the value zero into the calculator. Player B will begin play by selecting a number from List E (evens) or List O (odds). If the number is chosen from List E, it is added to the value in the calculator. If the number is selected from List O, it is subtracted from the value in the calculator. List E is the infinite list of numbers: $1/2, 1/4, 1/6, 1/8, \ldots, 1/2n, \ldots$ List O is the infinite list of numbers: $1, 1/3, 1/5, \ldots, 1/(2n-1), \ldots$ Player A then tells Player B whether or not the value in the calculator is greater than Player A's selected number, less than Player A's selected number, or equal to Player A's selected number. If the value in the calculator is equal to Player A's number, play terminates. If not, Player B selects another number *that has not been selected before*, from either List E or List O and adds it to the value in the calculator if it is from List E or subtracts it from the value in the calculator if it is from List O in an effort to get closer to Player A's number. Play continues until Player B determines Player A's number or until Player B has chosen 20 numbers from the lists. (Player B may not use any number from the lists more than once, but may choose numbers in any order.) In the latter case, Player B may make a final guess (without further calculations on the calculator).

Player B's score is 20 points if Player A's number is determined exactly; otherwise it is n points if the value guessed matches n decimal digits exactly.

Notes: Player B needs to keep track of those numbers that have been used. It is suggested that numbers be written down as they are used while play is in progress. It is also suggested that values in the calculator be written down (or stored in memory) as they appear, so that if a mistake is made entering a number the previous value may be recovered.

For example, a sample game where Player A chooses 0.42000000 (unknown to Player B) is given on the next page.

Play this game four times, with each partner assuming the role of Player A twice. Then answer the following questions.

1. What strategies were developed for Player B as the games were played?

Player B chooses	New calculator value	Player A responds
1/2	$0 + 1/2 = .5$	Too high
1/3	$.5 - 1/3 = .166\ldots$*	Too low
1/6	$.166\ldots + 1/6 = .3333\ldots$	Too low
1/10	$.3333\ldots + 1/10 = .4333\ldots$	Too high
1/15	$.4333\ldots - 1/15 = .3666\ldots$	Too low
1/30	$.3666\ldots + 1/30 = .40000$	Too low
1/60	$.40000\ldots + 1/60 = .41666\ldots$	Too low
1/100	$.41666\ldots + 1/100 = .42666\ldots$	Too high

Etc. (What number would you choose next?)

2. What strategies were developed by Player A to prevent Player B from determining Player A's number?

3. If the game were to continue "indefinitely," do you think that Player A's number could be determined exactly? Why or why not?

4. Once you get a successive combination of "too high, too low," or vice versa, can you give an upper bound on how far off Player B is?

Problems

1. Can you determine a number N and a strategy so that Player B can always determine Player A's number (which contains eight decimal places) within N calculator entries?

2. Suppose the rules were changed so that Player B could pick from either list, as before, but that the number that had to be chosen from the selected list was the first unused number. If play were to continue indefinitely, would Player A's number always be determined?

3. Develop an extension of this game by varying one or more of the rules.

This activity gives students a "feel" for series with oscillating terms. Time is well spent, because the results mentioned above are almost immediately obvious as a result of the players developing strategies. Moreover, most strategies bear a resemblance to the bisection method for determining a root of a continuous function, and, is a nice way to revisit that topic.

This activity can be used to illustrate how conditionally convergent series can be rearranged to converge to any preassigned number.

---

*The number 1/3 is subtracted because 1/3 is from List O.

Approximating functions with polynomials

Topic: Introduction to Taylor polynomials

Summary: Students find polynomials that satisfy "initial conditions" at 0. From this, they should essentially deduce the formula for Taylor series coefficients.

Background assumed: Derivatives

Time required: 15 minutes

Thread: approximation and estimation

---

1. Find a polynomial $p(x)$ with $p(0) = 2$ and $p'(0) = 3$.

   Students will have no difficulty here.

2. ... and with $p''(0) = 4$.

   Most students will get this. Have one explain if anyone does not.

3. ... and with $p^{(3)}(0) = -2$.

   This is now pretty straightforward.

4. If $p(x)$ is a polynomial, say $p(x) = a_0 + a_1 x + a_2 x^2 + \cdots + a_n x^n$, find $p^{(k)}(0)$, for $k = 0, 1, 2, \ldots, n+1$.

Suggest not doing multiplications along the way for students who are not seeing the pattern.

---

5. If you had a finite list of numbers, $c_0, c_1, \ldots, c_n$, can you find a polynomial $p(x)$, such that $p^{(k)}(0) = c_k$, $k = 0, 1, \ldots, n$?

This will not be clear to everyone.

Problem
Find a polynomial $p(x)$ such that $p^{(k)}(0) = k!$, $k = 0, 1, \ldots, n$. (Recall that $k! = k(k-1)\cdots 1$.)

The first three parts—indeed the entire activity—can be done on the first day of Calculus 2 as a way of reviewing derivatives and previewing one topic that will be central to Calculus 2.

Introduction to power series

Topic: Power series

Summary: Students derive Taylor coefficients by computing derivatives of an arbitrary polynomial.

Background assumed: None

Time required: 30 minutes

Thread: approximation and estimation

In many cases, it is useful to approximate a function $f(x)$ near some point (say x_0) by a polynomial $p(x)$. One example is the tangent line, which is a first-degree polynomial. Recall that the tangent line is the polynomial whose value and derivative both agree with the value and derivative of the function f at x_0. Letting $t(x)$ denote the tangent line, we have chosen $t(x)$ to satisfy $t(x_0) = f(x_0)$ and $t'(x_0) = f'(x_0)$. Not surprisingly, by taking a quadratic, it is possible to "match" the function, the first derivative, and the second derivative (at x_0). And, in fact, by taking an nth-degree polynomial, we can match the function and the first n derivatives. In this activity we will discover how to do that. To simplify things a bit, for now we will take $x_0 = 0$.

1. Let $p(x) = a_0 + a_1 x + a_2 x^2 + \ldots a_n x^n$. Find $p(0), p'(0)$, and $p''(0)$. Find $p'''(0)$. Can you guess $p^{(4)}(0)$? Check this.

2. For $j \leq n$, what is $p^{(j)}(0)$?

3. Suppose f is a function whose value and first 5 derivatives at $x = 0$ are 1. (Can you think of such a function?) What 5th-degree polynomial matches f and its first five derivatives at $x = 0$?

Problems

1. Let $f(x) = \frac{1}{(1-x)}$. Find the first three derivatives of f at $x = 0$. (It is probably easiest if you think of $f(x)$ as $(1-x)^{-1}$.) Guess the fourth derivative, and check that your guess is correct. What 4th-degree polynomial approximates $f(x)$ near $x = 0$? What n^{th}- degree polynomial approximates $f(x)$ near $x = 0$?

2. Let $f(x) = \cos x$. Find the first four derivatives of f at $x = 0$. What 4th-degree polynomial approximates $f(x)$ near $x = 0$? What n^{th}-degree polynomial approximates $f(x)$ near $x = 0$?

Graphs of polynomial approximations

Topic: Taylor polynomials

Background assumed: Students should have seen numerical series or power series, know the long division algorithm for polynomials, and know how to graph simple functions by graphing calculator or computer.

Time required: 50 minutes

Resources needed: Graphing software or graphing dalculator

Thread: approximation and estimation

In this activity, we will look for a way to approximate a non-polynomial function by a sequence of polynomial functions of higher and higher degree. If we choose the approximating polynomials carefully, each will match the original function more and more closely in the vicinity of a particular chosen point.

Let $f(x) = \dfrac{1}{1+x}$. We will construct approximating polynomials

$$P_0(x), P_1(x), P_2(x), \ldots \text{ where}$$

$P_0(x)$ is a constant function,

$P_1(x)$ is a linear function,

$P_2(x)$ is a quadratic function,

$P_3(x)$ is a cubic function,

$\vdots$

We will focus on the base point $x = 0$.

For this example, we can obtain the polynomials we want by performing a little algebraic "magic."

1. Divide the number 1 by the expression $(1+x)$, using the long division algorithm. Carry the algorithm out enough to yield at least seven terms of the quotient.

Some students will need to be reminded about the long division process for polynomials. Many will be a bit confused, because the terms in the dividend are arranged from lower to higher powers of x.

It's a good idea to check out their answers to the long division problem before they go much further, and to talk about the answer as a *power series*.

From this result, you can probably guess how the rest of the long division process would go. The quotient is an infinite series in powers of x, or a *power series in x*.

Depending on how much they already know about series, they might also be able to verify that the series is geometric, find its sum, and decide some values of x for which it converges.

Now, rather than focusing on the entire quotient, let's take only the first seven terms—which turns out to be a polynomial of degree six. We will use this polynomial for $P_6(x)$.

We started out with the idea of finding polynomials that would serve as good approximations of our original function $f(x) = \dfrac{1}{1+x}$. Let's see how good a job P_6 seems to do.

2. Use your graphing calculator or the graphing software on your computer to graph f and P_6 on the same set of axes. Graph both functions on the open interval $(-1, 1)$.

3. Does P_6 seem to be a good approximation for f near $x = 0$?

4. Describe in words why P_6 is or is not a close match for f near $x = 0$.

You will probably agree that it is important for us to decide just what we mean when we say that two functions "match closely" or that one is a "good approximation" for the other near the point $x = 0$.

To do this, we will look at the lower-degree polynomials that our division algorithm yields. That is:

$$\begin{aligned} P_0(x) &= 1 \\ P_1(x) &= 1 - x \\ P_2(x) &= 1 - x + x^2 \\ P_3(x) &= 1 - x + x^2 - x^3 \\ P_4(x) &= 1 - x + x^2 - x^3 + x^4 \\ &\vdots \end{aligned}$$

5. Graph f and P_0 on the same set of axes. Graph both functions on the open interval $(-1, 1)$. Use your graphing calculator or the graphing software on your computer.

6. Does P_0 appear to be a good approximation to f? What do the graphs of P_0 and f have in common? Do you think that any reasonable approximating polynomial ought to have this property?

7. Now graph f and P_1 on the same set of axes. Graph both functions on the open interval $(-1, 1)$. Use your graphing calculator or the graphing software on your computer.

8. Does P_1 appear to be a good approximation to f? What do the graphs of P_1 and f have in common? Do you think that any reasonable approximating polynomial ought to have this property? Do you believe this is the most appropriate way to approximate f near $x = 0$ *using a linear function*?

Let's look at your last few answers. You probably discovered that P_0 and f have the same value at $x = 0$. You probably also decided that it seems reasonable to insist that any good approximation have the same value at the base point, $x = 0$, as the function f. Since our first approximating polynomial had to be a constant function, the only possible choice of the constant was $f(0)$. That is:

$$P_0(0) = f(0).$$

9. Verify that P_1, P_2, P_3, P_4, P_5, and P_6 all take on the value $f(0)$, or 1, when $x = 0$.

Now let's look at P_1. You undoubtedly identified this linear function correctly as the tangent line to the graph of f at $x = 0$. So the slope of P_1 is the derivative of f at $x = 0$, $f'(0)$. Of course, the slope of a line is the (first) derivative of the corresponding linear function. So we see that $P_1'(0) = f'(0)$. The tangent line also passes through the point $(0, f(0))$. So the linear function P_1 and the original function f are related by two conditions:

$$P_1(0) = f(0) \text{ and } P_1'(0) = f'(0).$$

We'll continue to examine the approximating polynomials in a similar fashion.

10. Graph f and P_2 on the same set of axes. Graph both functions on the open interval $(-1, 1)$.

11. Verify that P_2 and f have the same value and the same slope at $x = 0$. I.e., $P_2(0) = f(0)$ and $P_2'(0) = f'(0)$. Describe an additional similarity between P_2 and f near $x = 0$.

The additional similarity, of course, has to do with concavity. Both graphs are concave up near $x = 0$. Concavity is determined by a function's second derivative.

12. Verify that $P_2''(0) = f''(0)$.

Problems

1. Graph f and P_3 on the same set of axes. Graph both functions on the open interval $(-1, 1)$. Verify that $P_3(0) = f(0)$, $P_3'(0) = f'(0)$, $P_3''(0) = f''(0)$, and $P^{(3)}(0) = f^{(3)}(0)$.

2. Compare the graphs of f and P_4 near $x = 0$. Find and verify *five* properties that f and P_4 share.

3. Generalize your answer to problem 2 to describe the nth approximating polynomial, P_n.

Taylor series

Topic: Taylor series

Summary: Students derive (and verify) a Taylor series for x^3 at 2.

Background assumed: Maclaurin series

Time required: 15 minutes

Resources needed: Show graphs of some of the partial sums for $\ln x$. The radius of convergence for this series is 1. It may come as a surprise that convergence "stops" at $x = 2$; it is a nice application of (or introduction to) the notion of radius of convergence.

Threads: approximation and estimation, multiple representation of functions

Comment: The introduction of power series expands the number of ways we can represent functions.

Recall that to construct a power series in x for a function f, the coefficient of x^n is

$$a_n = \frac{f^{(n)}(0)}{n!}.$$

It is also possible to construct a power series for f in $x - a$. In this case, the coefficient of $(x-a)^n$ is

$$a_n = \frac{f^{(n)}(a)}{n!},$$

and the resulting series is

$$\sum_{n=0}^{\infty} a_n (x-a)^n.$$

We will call this the power series for f at a. As a simple example, we will find the power series in for the function $f(x) = x^3$ at 2.

1. Find the power series for $f(x) = x^3$ at 2. That is, $a = 2$, and we are finding a series in terms of powers of $(x-2)$.

2. Check your work by multiplying out the $(x-2)^n$ terms and simplifying.

This is pretty straightforward. Students easily calculate and evaluate derivatives of x^3. Expanding $(x-2)^3$ will take a few minutes, and students are usually pleased to see that things work out because "everything goes away."

It is useful to assign at least one function that is undefined at zero so the students see why this (expanding about $x = a$) is important.

Problem
Find the power series for $f(x) = \ln x$ at 1.

Approximating logs

Topic: Taylor series

Summary: Students use Taylor series to approximate logs, including values for which the Taylor series doesn't converge.

Background assumed: Taylor series, properties of logarithms, alternating series

Time required: 20 minutes

Thread: approximation and estimation

---

1. Show that the Taylor series for $\ln x$ about $x = 1$ is given by

$$(x-1) - \frac{(x-1)^2}{2} + \frac{(x-1)^3}{3} - \frac{(x-1)^4}{4} \pm \cdots \qquad (*)$$

2. Show that $(*)$ converges for $0 < x \leq 2$. In particular, note that the series diverges for $x > 2$.

3. Use the Taylor series to approximate $\ln 1/3$.

4. Use the estimate in 3 to approximate $\ln 3$.

5. How accurate is your estimate of $\ln 3$? (Use a calcalculator to check.)

6. In general, how can one use $(*)$ to estimate $\ln x$ for $x > 2$?

Problem

1. Estimate $\ln 5$.

Using series to find indeterminate limits

Topics: Limits, power series

Summary: This activity gives the students an application of series—to find indeterminate limits.

Background assumed: Power series for sine and cosine.

Time required: 15 minutes

Thread: approximation and estimation

You may (or may not) recall the value of $\lim_{x \to 0} \dfrac{1 - \cos x}{x}$. Here we will use power series to find it.

First, recall that the power series for $\cos x$ is $1 - \dfrac{x^2}{2!} + \dfrac{x^4}{4!} - \cdots$. So,

$$1 - \cos x = \frac{x^2}{2!} - \frac{x^4}{4!} + \cdots.$$

Dividing by x, we have

$$\frac{1 - \cos x}{x} = \frac{x}{2!} - \frac{x^3}{4!} + \cdots;$$

so it follows that

$$\lim_{x \to 0} \frac{1 - \cos x}{x} = 0.$$

This idea is easy enough: compute the power series for the numerator and denominator, divide, and look at the limit.

1. Find $\lim_{x \to 0} \dfrac{1 - \cos x}{x^2}$.

2. Find $\lim_{x \to 0} \dfrac{1 - \cos x}{x^3}$.

3. Find $\lim_{x \to 0} \dfrac{\sin x}{x}$.

4. Find $\lim_{x \to 0} \dfrac{\sqrt{1 - \cos x}}{x}$. (Hint: How does this problem compare to 1?) (Note: Try l'Hôpital's rule. Can you figure out a way to obtain this limit using l'Hôpital's rule?)

It is also instructive to give an example with an infinite limit (but not indeterminate) to see what series is obtained. For example, $\lim_{x \to 0} \dfrac{2 - \cos x}{x}$.

The last problem is difficult for students using l'Hôpital's rule. One use of l'Hôpital's rule yields the limit $\lim_{x \to 0} \dfrac{\sin x}{2\sqrt{1 - \cos x}}$, which can be computed by multiplying the numerator and denominator by $\sqrt{1 + \cos x}$. (Most students will not think of this!) A second application of l'Hôpital's rule gives $\lim_{x \to 0} \dfrac{\cos x \sqrt{1 - \cos x}}{\sin x}$, which is (almost) the same problem again. (Factor out $\cos x$, since its limit is 1.) It's informative to try to have the students use the fact that this limit is twice the limit of its reciprocal to find both.

This is a nice example to illustrate that series are often easier to use than l'Hôpital's rule. Without resorting to the first example, one can note that if $1 - \cos x \approx \frac{x^2}{2}$, then $\sqrt{1 - \cos x} \approx \frac{x}{\sqrt{2}}$.

Using power series to solve a differential equation

Topics: Differential equations, power series

Summary: Students develop a technique to solve differential equations using power series.

Time required: About 20 minutes

Background assumed: Students need to know the power series for e^x and what a differential equation is.

Resources needed: If software is available, plot the successive solutions. If it is convenient, show the direction field.

Thread: approximation and estimation

Our goal in this activity is to find a power series ("infinite polynomial") that solves a differential equation. We will work specifically with the differential equation $y' = 2y$ satisfying the initial condition $y_0 = 1$ (which means that $y = 1$ when $x = 0$ or, more simply, $y(0) = 1$). Since you know the solution is given by $y = e^{2x}$, this will also serve as a derivation for the power series for e^{2x}. However, nothing we do here depends on our knowing the solution beforehand; indeed, this technique can be used to find a solution to a differential equation that can't be solved in other ways.

Problem: Find a solution to $y' = 2y$, $y_0 = 1$.

1. Begin with a constant polynomial, $p_0(x) = c_0$. Be sure it satisfies the initial condition.

2. Now construct a polynomial, $p_1(x)$, whose derivative is $2p_0(x)$. Be sure it satisfies the initial condition.

Students will not have difficulty with the first few parts, but the idea behind this technique needs comment.

The basic idea is that higher degree polynomials should approximate better. So we start with p_0, a constant function—all it can do is satisfy the initial condition. We then construct p_1, a linear function that satisfies the differential equation if p_0 is y (on the right-hand side). And so on. This idea, using p_n on the right-hand side to produce a p_{n+1}, may be confusing.

---

3. Next, construct a polynomial, $p_2(x)$, whose derivative is $2p_1(x)$. Be sure it satisfies the initial condition.

Suggest that they *not* multiply things out if they have a hard time seeing the pattern.

---

4. Next, construct a polynomial, $p_3(x)$, whose derivative is $2p_2(x)$. Be sure it satisfies the initial condition.

5. What series solves the differential equation $y' = 2y$, $y_0 = 1$?

This example is so "clean" that students may not realize what the technique is; it is helpful to do one of the follow-up problems to help make that clear. The third problem below is particularly nice, since the students probably do not know any other technique to find a solution. It only takes a few steps before the pattern becomes clear. However, students will have difficulty identifying the series solution as the sum of an exponential and a linear function. It is probably best not to have students dwell on this, although it is valuable for the students to verify that the function is indeed a solution. (One way to avoid the algebra is to give the closed-form solution and ask the students to verify that its power series is the one they found as the solution to the differential equation.)

Problems

1. Find a series solution to $y' = 2y$, $y_0 = 2$. What function does this series represent?
2. Find a series solution to $y' = y$, $y_0 = 1$. What function does this series represent?
3. Find a series solution to $y' = x + y$, $y_0 = 1$. What function does this series represent?

The solutions are $y = 2e^{2x}$, $y = e^x$, and $y = 2e^x - x - 1$.

Second derivative test

Topics: Taylor polynomials, second derivative test

Summary: Students are led to see the relationship between the power series for a function and the function's derivatives. This should lead to new insight to the second derivative test.

Time required: 10–15 minutes

Background assumed: Power series

Thread: approximation and estimation

Suppose all the derivatives of g exist at 0 and that g has a critical point at 0.

1. Write the nth Taylor polynomial for g at 0.

The linear coefficient, a_1, is zero.

---

2. What does the second derivative test for local maxima and minima say?
3. Use the Taylor polynomial to explain why the second derivative test works.

We have $g(x) = g(0) + a_2 x^2 + x^3(a_3 + \ldots)$. For small x, the x^2 term is dominant, so if $a_2 > 0$, $g(x) > g(0)$ (x small). (Be prepared to explain this point!) Since $a_2 = g''(0)/2!$, we see that $g(0)$ is a local minimum if $g''(0)$ is positive. Similarly, $g(0)$ is a local maximum if $g''(0)$ is negative.

Problem
   Suppose in addition that $g''(0) = 0$. What does the Taylor polynomial tell you about whether g has a local maximum or minimum at 0?

Now, in addition, $a_2 = 0$. If $a_3 \neq 0$, then (similar to the argument above) $g(x)$ will be greater than $g(0)$ on one side of zero and less on the other (depending on the sign of a_3); thus g does not have a local extremum at 0.

Ask the students what happens if $a_3 = 0$. Suggest that they think about what happens if $a_1 \neq 0$, what happens if $a_1 = 0$ and $a_2 \neq 0$, what happens if $a_1 = a_2 = 0$ and $a_3 \neq 0$. Mention the importance of the first (ignoring a_0) non-zero term in determining the local behavior.

In general, if the first non-zero term of the Taylor polynomial corresponds to an even power of x, then there is a local extremum at 0 (minimum if a_{2m} is positive, maximum if a_{2m} is negative); if the first non-zero term corresponds to an odd power of x, then there is no local extremum at 0.

Padé approximation[2]

Topic: Padé approximation

Summary: Students find an approximation of e^x with rational functions.

Background assumed: Students should have seen Taylor polynomials.

Time required: 40 minutes

Resources needed: Graphing software or a graphing calculator is necessary for question 4. (If not available to the students, the instructor can bring in these graphs.) If a computer algebra system is available, it may be used to assist with the calculations.

Thread: approximation and estimation

---

[2] Adapted from Deborah Hughes-Hallett, Andrew Gleason, *et al.*, *Calculus*, John Wiley, 1992.

In this problem, you will find a rational approximation to the exponential function. (Such approximations are called Padé approximations.)

1. Let $f(x) = \dfrac{a + bx}{1 + cx}$, where a, b, and c are constants. Find $f(0)$, $f'(0)$, and $f''(0)$.

These calculations are straightforward, but messy: $f(0) = a$, $f'(0) = b - ac$, and $f''(0) = -2c(b - ac)$. Make sure students have these correct before proceeding!

---

2. Let $g(x) = e^x$. Find $g(0)$, $g'(0)$, and $g''(0)$.

This should be easy; all of these are 1.

---

3. By equating the values of you found (i.e., making f and g have the same function value, first derivative, and second derivative at 0), find a, b, and c so that $f(x)$ approximates e^x as closely as possible near $x = 0$.

Finding the value of a is easy. It is convenient to note the relationship between $f'(0)$ and $f''(0)$ as a way of determining c. The solution is $f(x) = \dfrac{1+x/2}{1-x/2}$.

---

4. Graph $f(x)$ and e^x on the interval $[-3, 3]$.
5. Comment on the value of using this method to approximate e^x.

The approximation is fine near 0. It fails badly where f blows up.

Problems

1. How would you find the Padé approximation for $f(x) = \cos x$?
2. How would you find the Padé approximation for $f(x) = \sin x$?
3. How could you find a more accurate Padé approximation for e^x?

One can improve the approximation by taking higher-order terms in the numerator and denominator.

It's nice to compare graphs of e^x, the Padé approximation, and various power series.

Using Taylor polynomials to approximate integrals

Topics: Integration, Taylor series

Summary: This activity uses Taylor polynomials in two different ways to approximate integrals, in one case by approximating the integrand, in the other by approximating the antiderivative of the integrand.

Background assumed: Taylor polynomials

Time required: 30 minutes

Resources needed: A calculator is useful.

Thread: approximation and estimation

In this activity, you will examine and compare two methods for approximating ln 2 using Taylor polynomials. The methods are useful in estimating the value of many definite integrals where one does not know an antiderivative for the integral or—even if an antiderivative is known—if the needed values of the antiderivative are not easily obtained.

Assume your calculator has a broken $\boxed{\ln x}$ button. That is, assume we need to estimate ln 2 by estimating the value of $\int_1^2 \frac{1}{x}\,dx$, and we only have elementary operations of arithmetic available. We will use two methods.

Approximate the integral Let $F(x) = \int_1^x \frac{1}{t}\,dt$. Find the fifth-degree Taylor polynomial for F at 1. Call this polynomial P_5. Now, estimate $F(2)$ by computing $P_5(2)$.

Approximate the integrand Let $f(t) = 1/t$. Compute the fifth-degree Taylor polynomial for f at 1. Call this polynomial p_5. Now estimate $\int_1^2 \frac{1}{t}\,dt$ by computing $\int_1^2 p_5(t)\,dt$. Do this by determining an antiderivative for p_5 and using the fundamental theorem of calculus.

After these two estimates have been computed, answer these questions.

1. What is the significance of using a Taylor polynomial at 1?

2. Which of these estimates is more accurate?

3. Describe the difference that results in these two methods.

4. In general, would you expect one of these methods to be a superior method for estimating the value of an integral?

Problems

1. Compile a list of methods that one can use to estimate $\int_1^2 \frac{1}{t}\,dt$.

2. Let p_2 be the second-degree Taylor polynomial at 1 for the function $f(t) = 1/t$. Let q_2 be a quadratic polynomial that fits the curve $y = 1/t$ at $(1,1)$, $(3/2, 2/3)$, and $(2, 1/2)$. Are p_2 and q_2 the same? Which will give you a better approximation of $\int_1^2 \frac{1}{t}\,dt$ when you integrate?

3. We will estimate $\int_1^2 e^{t^2}\,dt$ using a variation of the second method. Let $f(t) = e^t$. Find the fifth-degree Taylor polynomial, P_a, at a. Now, since $f(t) \approx P_a(t)$ for t close to a, $f(t^2) \approx P_a(t^2)$ for t^2 close to a^2. Decide on a good choice for a and defend your choice. Now, compute $\int_1^2 P_a(t^2)\,dt$. Can you give an estimate for your error? Finally, use the Taylor series for e^t and integrate to give a series that converges to $\int_1^2 e^{t^2}\,dt$.

This activity incorporates applications of Taylor polynomials and relates to the fundamental theorem of calculus, directly in the first method and indirectly in the second (and the homework)—in a case when you may not be able to find an antiderivative of a function. This is especially effective if $\ln x$ has been defined as $\int_1^x 1/t\, dt$; that is, $\ln x$ has been defined as an antiderivative of $1/x$ (for $x > 0$). During one course we taught, we frequently estimated $\ln 2$ as an application of various methods of approximation: Riemann sums, polynomial (quadratic) curve fitting to approximate an integral, etc. If other methods have been used to approximate $\ln 2$, they should be compared with the results from this activity. This activity could precede a discussion of error estimates with Taylor polynomials and Taylor series (omit the last parts of problem 3 in this case) and could be used to motivate such a discussion.

Complex power series

Topics: Complex numbers, power series

Summary: The students are led to develop and use Euler's formula.

Background assumed: Students should know the Maclaurin series representations for $\sin x$, $\cos x$ and e^x.

Time required: 50 minutes

Thread: approximation and estimation

A complex number can be thought of as an ordered pair of real numbers (x, y)—which we will write as $x+iy$—where i is defined as the square root of -1. We call x the *real part* of the complex number $x+iy$, and y is called the *imaginary part* of the complex number. We can add, subtract, multiply, and divide complex numbers by treating $x + iy$ as an algebraic expression with the rule that $i^2 = -1$. For example,

$(3 - 4i) + (6 + 2i) = (3 + 6) + i(-4 + 2) = 9 - 2i$

$(3 - 2i)(5 + i) = 15 - 10i + 3i - 2i^2 = 15 - 7i - 2(-1) = 17 - 7i$

1. Find the following products:

    (a) $(4 + 2i)(4 + 2i)$
    (b) $(\cos\theta + i\sin\theta)(\cos\theta + i\sin\theta)$
    (c) $(ix)^3$, $\quad x$ real
    (d) $(ix)^4$, $\quad x$ real

2. Write the power series expansion (Maclaurin series) for e^x.

3. In the expansion for e^x, substitute $x = i\theta$ where θ is real.

4. Your expansion will involve some terms that include i or $-i$ and other terms that only involve real numbers. Group all the terms that only involve real numbers and then group all the terms that involve i or $-i$.

5. The result of 4 should be two power series. The first power series only involves real numbers and the second power series has i or $-i$ in each term.

    (a) Factor i out of the second power series and write it as i times a power series that only involves real numbers.

    (b) Write the result as

    (real power series in powers of θ) + i (real power series in powers of θ).

    (c) You should recognize each of the power series in 5b as a well known function of θ. Identify these two functions.

    (d) When you substitute these functions into 5b you obtain a formula for $e^{i\theta}$, called Euler's formula. Do the substitution and write down Euler's formula.

6. Euler's formula can be used to help you remember some trig identities.

    (a) First, using the properties of exponents show that
    $$e^{i\theta} e^{i\theta} = e^{i2\theta}.$$

    (b) Next, replace each copy of $e^{i\theta}$ by Euler's formula and multiply these two complex expressions.

    (c) Finally, use Euler's formula to replace $e^{i2\theta}$ by a complex expression.

    (d) Complex expressions are equal if and only if their real and imaginary parts are equal. Use this fact to obtain a formula for $\cos(2\theta)$ and another formula for $\sin(2\theta)$.

7. Use the formulas in the previous step to find $\cos^2\theta$ and $\sin^2\theta$ in terms of $\cos(2\theta)$.

8. Use the formulas in the previous step to integrate $\cos^2\theta \, d\theta$ and $\sin^2\theta \, d\theta$.

The instructor can do parts 6, 7, and 8 or assign these parts as homework in order to save time.

Part II

Projects

Calculus 1   Project 1

Designing a roller coaster

Background assumed: Increasing and decreasing functions, concavity

Summary: This project uses the concepts of increasing, decreasing, and concavity to analyze the path of a roller coaster. Students construct graphs of the first and second derivatives of the path of the roller coaster and can use these graphs to find the steepest part of the path. They also relate the slope of the path to the angle the path makes with a horizontal line. Other mathematical concepts include constructing a curve that satisfies given constraints and then optimizing curves with respect to certain conditions.

Threads: graphical calculus, modeling, approximation and estimation

Objectives:

1. Working with graphs (recognizing increasing, decreasing, and concavity)

2. Relating slope and angle

3. Constructing graphs of first and second derivatives from given graph

4. Modeling and optimizing

5. Using slope graph to find max or min (discovery approach)

Prerequisites:

1. Increasing, decreasing and concavity for graphs

2. Constructing slope graphs from a given graph

Notes:

1. This project gives students an outline to follow for much of the problem, but it does include some open-ended pieces.

2. Each student was given their own design to analyze for part A. We have included the individual designs for one group.

3. Our experience is that the terms decreasing at an increasing rate and decreasing at a decreasing rate can cause confusion. You can discuss these terms in class or define them for the class. If you wish to avoid this type of discussion replace the terms increasing at an increasing rate, etc., by increasing and concave up, etc.

4. The definition of thrill is arbitrary. You can replace it if you wish. The given definition implies that a coaster with several small hills (a "kiddie coaster") is more thrilling than a roller coaster with a few long steep drops. As a follow-up you could ask the class to find definitions of thrill that make coasters with steep drops more thrilling than coasters with several small drops. You also can briefly discuss the idea of "metrics" for families of curves and the calculus of variations when you return the projects.

5. Be sure to point out that the the maximum point on the graph of the slope corresponds to places where the rate of change of the slope changes from positive to negative. This is using the first derivative test to find maximuma of f'.

6. In part C, question 1, students are expected to estimate the distance traveled.

7. Here are some other coasters we have given to students.

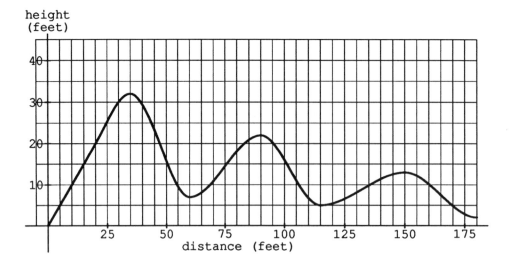

Coaster A

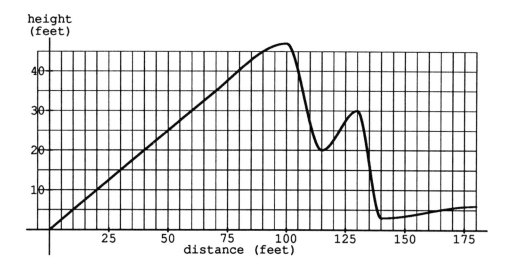

Coaster B

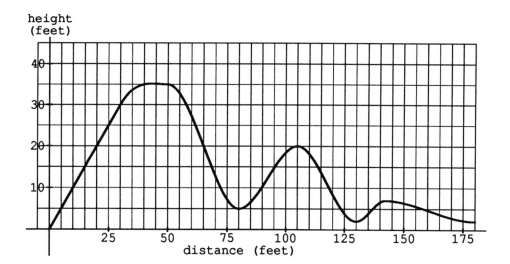

Coaster C

Calculus 1 — Project 1, Part 1

Tidal flows

Topics: Rate of change, preview of the fundamental theorem of calculus.

Summary: Part 1 is intended to introduce rate of change as slope, in a context other than distance-velocity. Part 2 introduces the idea of using Riemann sums to estimate aggregates from a rate graph.

Background assumed: Students should be comfortable with the idea of representing a function just as a graph, and with the idea of estimation from a graph.

Time required: About 2 weeks, as a group project, for both parts.

Threads: graphical calculus, approximation and estimation, modeling

Comment: In part 2,

1. The way to construct the graph of the total net flow is to add the two given rates graphically. There are several times when the tidal influence is great enough to cause water to flow "up river." Students need to realize that this effect corresponds to a negative flow rate.

2. Students doing this project early in the first semester will probably decide to calculate the volumes using Reimann sums and intervals of one day.

Calculus 1   Project 1

Designing a cruise control

Time required: Two weeks

Summary: This project is designed for early use in Calculus 1. It serves as an application of the concepts of composition of functions, piecewise-defined functions, and graphical analysis of functions. The project also serves to motivate the concepts of limits and continuity. It also illustrates a top-down approach in that it breaks the overall problem into parts. Students are forced to see that a function is a mathematical "device" that determines a unique output for a given input.

Prerequisites: Functions as represented by graphs and formulas; composition of functions.

Threads: distance and velocity, graphical calculus, multiple representation of functions, approximation and estimation, modeling, top-down methodology

Comments:

1. In part 1, students can derive a formula for a function using a piecewise definition or the greatest integer function. The graph should be a step function.

2. In 2, the domain is discrete.

3. Note that in 3 the composition is a step function, not the identity as would be desired.

4. The "problem" in 4 is that the display would never read 55 mph, since that speed does not correspond to an integer number of clicks per second.

5. For 5, students should look at the corresponding graphs of functions they get when the number of equally space pins increases. These step functions converge uniformly *but not monotonically* to the identity. For example, if the car were actually traveling at 55 mph, 12 pins would give a more accurate display than 13 pins! Nonetheless, students should believe that as the distance between pins approaches zero, the display's speed would approach the actual speed of the car. Later, when discussing sequences this could be expressed by saying that the limit, as the number of pins approaches infinity, of the displayed speed is the actual speed of the car. Transmitting the count faster would make the display less accurate.

6. Part 6 could be used to motivate the idea of a Riemann sum: the areas under the step functions generated from additional pins are all lower sums for the function $f(x) = x$.

Calculus 1 — Project 2

Designing a detector

Topic: Periodic functions

Summary: This is an example of an open-ended project. Students model the position of a detector both graphically and analytically. This project illustrates the mathematical modeling process very well. The initial "solutions" will go through a number of refinements as students modify their assumptions. Students also encounter a mathematics problem where there is no "correct" solution.

Background assumed: Derivatives, relationship between position and velocity

Time required: About two weeks

Threads: modeling, multiple representation of functions, graphical calculus, distance and velocity

Comments:

1. Part A. Many groups will make x a piecewise-linear function of t. This lets them keep the speed of the detector constant as it moves up and down the hall. A problem with this approach is that the graph has a corner each time the detector reaches the end of the hall. So there are corners in graphs defined like this at the points where $x = \pm 20$ since the velocity changes from s (the speed of the detector) to $-s$ at such points. This means the detector will be bumped at the end of every trip down the hall. Discussion of this problem will lead most groups to consider rounding off the corners (that is, consider questions of differentiability). This usually led the groups to consider graphs like the sine curve.

2. Part B. Groups that use the piecewise-linear function may have trouble finding the analytic form. Most of the rest will use a rule of the form $x(t) = 20\sin(kt)$. Be prepared to answer questions related to finding the value(s) of k so that the detector will go up and down the hall within a given time period. Also students may have difficulty in finding the derivative of $x(t)$ if chain rule has not been discussed yet. There may also be questions about inverting the sine function. These question arise when the groups try to determine the time the door is covered.

Calculus 1   Project 1

Taxes

Topics: Graphical differentiation and integration

Summary: The most important aspect of this project is the relationship between the tax function and the marginal tax function. Another important theme is multiple representation of functions; functions arise in algebraic and graphical contexts throughout. Students have to use and construct some piecewise-defined functions. There are several open-ended aspects.

Background assumed: Students must have seen the relationship between a function and its "rate graph" and "area graph."

Time required: About two weeks

Threads: multiple representation of functions, graphical calculus

Comments:

1. Part A is essentially a warm up intended to get the students to understand the three different "tax graphs" (the tax, average tax, marginal tax) and to get accustomed to working together. Part B is also straightforward, although some students will not see how to turn their social security number into a tax function. (In particular, some students will not see where the number 32,250 in the example comes from. Do not tell them that it is 0.75 times 43,000; rather, ask them to graph the function. If they still do not recognize its significance, ask what the graph would look like if 32,250 were replaced by some other number. They should realize that it is the number that makes the function continuous.) Parts A and B should be due in about one week.

2. Most or all groups will realize that the marginal tax is the derivative ("slope graph") of the tax (and vice versa, tax is the "area graph" for marginal tax), and obtaining information about the average tax from tax is easy. Working *from* the average tax is more difficult; suggest that they think about the relationship between tax and average tax.

3. In Part D, students usually use the midpoint of the income range as the basis for their estimate. This does not work for the unbounded range, and students should be encouraged to explain how they deal with this. (Some groups

stipulate a largest taxable income. These are sometimes unrealistically low, in which case using the midpoint for that range may be appropriate; if their maximum is a nine-figure income, estimating with the midpoint is probably not reasonable.)

Calculus 1 — Project 2

Water evaporation

Topics: Derivatives, differences, functional equations

Background assumed: Increasing and decreasing functions, concavity

Summary: This project requires students to work with a function given in tabular form. It also involves relating the tabular form to graphical and algebraic properties.

Resources needed: None needed. However, a spreadsheet can be used to compute the table of differences.

Thread: multiple representation of functions

Comments:

The given function is $25e^{-.2x}$. We have given each group different data by simply using $25e^{kx}$ with values of k such as $-.1, -.2, \ldots$

In part A, question 5, be sure that the class knows that the slope of f is equal to $-.2$ times f. This will be useful when you get to the chain rule.

Calculus 1   Project 4

Mutual funds

Topic: Fundamental theorem of calculus

Background assumed: Multiple representation of functions, rates of change, increasing and decreasing functions.

Summary: This project focuses on the representation of functions via tables, algebraic expressions, verbal descriptions, and graphs. Students are asked to write several functions in each of the three ways. The relationship between a rate of change and total value is explored. While this hints at the fundamental theorem of calculus, this project may be done near the beginning of Calculus 1 before this theorem is mentioned or along with an initial discussion of the theorem. Questions 6, 7, 8, and 9 have the students comparing rates of increase and decrease, considering how these rates are affected under basic addition of functions, and optimizing functions early in the course.

Time required: About two weeks

Threads: multiple representation of functions, graphical calculus

Comments:

1. In parts 6, 7, 8, and 9 you can transfer your money back and forth between funds as often as you wish with no penalty.

2. Plotting the rate of change of each fund on the same set of axes is an easy way to see how to solve parts 6, 7, and 9.

Calculus 1   Project 4

Rescuing a satellite

Topics: Integration, graphical integration

Background assumed: Riemann sums, integrals as area under a curve

Summary: Students use Riemann sums to estimate integrals and decide whether a rocket ship will catch a satellite within two years.

Threads: multiple representation of functions, distance and velocity, top-down methodology. approximation and estimation

Comments:

1. This is a nice application of the intermediate value property.

2. Some groups may have trouble with the units. Both velocities are given in miles per hour after t years. The graphical approach (distance traveled is area under the velocity curve) illustrates the difficulty. One block would have an area of $0.25 \cdot \frac{1}{12}$ thousand-miles-per-hour years.

3. The rescue ship will catch the satellite before the two years is up.

Calculus 2 — Project 2

Spread of a disease

Topics: Differential equations, iteration

Summary: Students compare itertion with the use of a differential equation to model spread of a disease. They examine the role of the constants in each model, and look for long run equilibrium.

Background assumed: Solution of differential equations by separation of variables.

Time required: About two weeks, for all parts.

Resources needed: A computer or calculator could be programmed to provide a way for students to explore further the effects of changing the constants in each model.

Threads: modeling, approximation and estimation

Objectives:

1. To have the students solve a number of separable differential equations on their own. Very little class time (less than one period) was spent on this before the project was assigned.

2. To explore the effects of changing parameters on the behavior of the systems.

3. To compare iteration with analytic (continuous) solutions of the model. This ties in with the emphasis on numerical approximation techniques that we have taken in the course and will be followed by a brief discussion of computer simulation as a solution method for dynamic systems.

4. To explore long-run behavior of the systems.

Comment: Note that Model 3 is logistic growth, and is a realistic model. Depending on the background of the students at the time the project is assigned, some instructors may wish to provide, or develop in class, the solution to the differential equation in this case.

Questions:

1. Is there any relation between k and c? More generally, suggest how the constants in the differential equations and the iterations are related. That is, given a differential equation model and its constants, how can an iteration be specified that approximates the differential equation. In which of the given models are the parameters c and k the only ones that need to be related? In which must other parameters be adjusted, and how?

2. For each model, determine parameter levels that lead to the disease being controllable under the scenario of the possibility of a cure. What does this imply about how accurately the parameters need to be estimated in order to make the model useful?

Calculus 2   Project 2

Tax assessment

Topics: Graphical and numerical integration

Background assumed: Integration

Summary: This project requires students to use integrals to calculate the value of various lots and to find how to equitably divide these lots.

Threads: approximation and estimation, modeling

Comments:

1. You can use different worth functions for different groups if you wish. If you do use different functions for each group you can illustrate the advantage of solving problems with parameters when you discuss solutions. For example, if $w(x) = A + Bx$, what values of A and B make sense and what will happen to the fair division point as A and B change?

2. When we assigned this project we gave each group an actual lot from a tax map and said it was assesed for $200,000. We then asked them to write a letter to the assessment board either supporting or protesting this assesment. The letter had to include a discussion of the method used for finding the value of the lot.

3. In part C some students use the average value of a square foot for a rectangular lot from part B and multiplying this number by the area of the trapezoidal lot. You can illustrate the problem with this approach by asking the students whether a lot with the same area but longer frontage should be worth more.

4. The $pf(x)$ function can be referred to as an example of a cumulative distribution function if you cover continuous probability in your course.

5. We gave each group a lot from an actual assessment map in our local area and told them that the lot was assessed for $500,000. The students were then asked to write a letter to the assessor supporting the claim that the assessment was too high. We told them the letter should include a description of the worth function they used, an explanation of why it is a "good" worth function, and how they used the worth function to determine the value of the lot.

Dome support in a sports stadium

Topic: Applications of integration

Background assumed: Volume computed using integration—at least one way using cross-sectional areas or solids of revolution. Also, surface area as an integral is recommended. Some plane geometry using right triangles (force resolution) and construction of a spherical cap of known altitude and base diameter.

Summary: This project supports several geometric applications usually discussed at the beginning of a second semester of calculus. Specifically, it involves computing the volume and weight of vertical buttresses. It also involves computing the surface area and weight of a spherical cap—the dome roof. Both of these involve integration. (A formula for the surface area of a spherical cap may be given to the students if surface area is not covered in the course.)

Time required: Two and one half weeks

Threads: modeling, top-down methodology

Comment:

1. The project is very open-ended in terms of the aesthetic design of the buttresses, and students should be encouraged to be creative. They also have freedom in choosing parameters for the spherical cap that forms the roof. Different choices of parameters will affect the height of the roof and the angle at which the roof meets the buttresses.

2. Their designs and choices need to create a structure that can pass certain structural tests: each buttress footer must be able to carry the weight of the buttress, a portion of the roof weight, and an additional weight component resulting from the maximum height of the stadium roof. In addition, the minimum cross-sectional area cannot be too small. The dome roof must cover a ground area with a diameter of at least 540 feet. (This is called the "span" of the roof.) The height of the roof, which depends both on the height of the buttresses and the parameters for the roof top affects the additional weight factor on each footer. The angle at which the roof meets the buttress also impacts on the minimum cross-sectional area of each buttress.

3. This project highlights the need for a good problem-solving strategy. A top-down approach might first break the problem into two sub-problems: the design of the buttresses and the design of the roof. Each of these can be further analyzed in terms of the criteria for meeting the specifications. The two major designs need to be interrelated, since both designs affect each structural test. It probably will take students some time to set up their approach. (You may want to ask students to present their strategy after two or three days of work on the project.)

4. Many students seem to have developed a real sense of accomplishment with the completion of this project. Students have provided carefully drawn sketches of their designs. Several considered a column in the shape of a solid of revolution where the graph which was rotated was not the graph of a function given by an algebraic formula. (I.e., the function was only created graphically.) Others have added factors including economic considerations to support their designs. Still other students have suggested modification of the design such as a dome in the shape of a parabaloid—with full confidence that they could do the work necessary.

Calculus 2 — Project 2

The fish pond

Topics: Differential equations, numerical integration

Background assumed: Integration (including numerical integration), separable differential equations

Summary: Students estimate the volume of water in a pond from a table of depth values. They use the volume to decide how much salt is in the pond by solving differential equations.

Threads: modeling, approximation and estimation, multiple representation of functions

Comments:

1. One of our report guidelines is to assume the reader is a student in a different calculus class. To decide whether or not another student could understand your report try this criterion: Given a different set of depth measurements, could the student compute the volume of the pond by reading your report?

2. Some groups mentioned the problem of whether the bottom had hills or valleys etc. One way to get an idea of this is to look at the second derivative. If all that is available is a table of values, people use second differences to approximate the second derivative. The second difference in the horizontal direction at a point x is given by $\frac{f(x+h)-2f(x)+f(x-h)}{h^2}$. In this case h would be 20. So the second horizontal difference at D2 is $\frac{2-2(10)+18}{20^2} = 0$, the second vertical difference at D2 is $\frac{12-2(10)+16}{400} = \frac{8}{400} = 0.02$, and so on. If all second differences are small then use of the trapazoidal rule is justified.

3. There are several different approaches that student groups used. We did not teach integrals of several variables in this course, but most groups did some version of finding a cross-sectional area and integrating the cross section to get the volume.

Calculus 2 — Project 2

Drug dosage

Topics: Exponential functions, partial sums and convergence, geometric series.

Summary: This project deals with models of drug concentration levels. It introduces numerical series and the concept of convergence.

Background assumed: Familiarity with the exponential function, and ability to solve the differential equation $y' = -ky$.

Time required: About two weeks for all parts.

Threads: modeling, approximation and estimation

Comments:

1. This project introduces numerical series in a concrete context. The students use graphs to learn about the ideas of convergence and divergence by examining the concentration levels in the blood of a drug that has been administered repeatedly.

2. The case of linear decay makes the student aware that divergence of the concentration level is possible, but that in some cases one might safely repeat doses indefinitely.

3. The linear case sets the stage for exponential decay where the series always converges, though not necessarily to a safe concentration level.

Calculus 2 — Project 3

Investigating series

Topics: Sequences, series, and Taylor polynomials

Summary: This project emphasizes pattern recognition and exploration with sequences and series. It also has students work with multiple representations of sequences, namely, graphical and numerical representations. The project works well as an introduction to series with students who have been introduced to sequences and are familiar with the concept of convergence. The idea in Part A is to represent the sequence b_n as a partial sum of the sequence $\{a_1, a_2, \ldots\}$. By viewing a sequence of partial sums in "closed form" as the sequence $\{b_1, b_2, \ldots\}$, students should begin to believe that a sequence of partial sums can converge, even though more and more summands are being included. Part B, formulated as a discovery activity, could be a harbinger of the relationship of series to the value of functions. Or, it could be reinforcement of material on Taylor and Maclaurin polynomials that the students have had. Part C gives the students experience with an alternating series which would help clarify a theoretical discussion of the convergence of alternating series.

Background assumed: Convergence of sequences, possibly Taylor and Maclaurin polynomials.

Time required: About two weeks, for all parts.

Threads: multiple representation of functions, approximation and estimation

Calculus 2   Project 4

Topographical maps

Topics: Level curves (intuitively), partial derivatives, directional derivatives (intuitively).

Summary: This project is intended to introduce visualization of surfaces, partial derivatives and directional derivatives, as a preview of vector calculus.

Background assumed: Derivative as a rate of change.

Time required: About one week.

Thread: graphical calculus

Comments:

1. In attempting to sketch what they can see from their vantage point, students need to realize that they cannot see objects in the landscape that are, for example, behind hills.

2. In estimating the rate of change in their altitude, they need to distinguish between y, the distance north, z, the altitude, and s, the distance traveled.

Part III

Appendices

Appendix A

Sample Curriculum

Calculus 1

We believe that calculus is a unified subject, not just a sequence of topics; consequently there are many different paths through a calculus course. We are presenting one sample curriculum which includes topics correlated with possible activities and projects from *Calculus: An Active Approach with Projects*.

Our goal is to illustrate how to have an active class that works on projects and activities without sacrificing content. We stress that this is only a brief outline. Hopefully, you can use this sample as a basis to design a path through calculus that fits your needs. Many of the activities can be used in a variety of places since they serve multiple roles. For each topic we have listed some activities that we have found work well for this particular choice of course organization. There are many other possibilities, however. In many cases, we have listed alternative activity choices.

Week 1

Introduction to modeling and graphical calculus

> Activities: Chalk toss, Classroom walk, Raising a flag. (Other possibilities include: Airplane flight with constant velocity, Projected image, Water balloon.)

Distance-velocity pair for linear distance case

> Distance as area under velocity
>
> Velocity as slope of distance
>
> Slope of a line and slope as rate of change
>
> Activities: Examining linear velocity, or More airplane travel

Finding rate of change from graph

> Activity: Graphical estimation of slope

Concepts of increasing, decreasing, concave up, concave down, and optima using distance and velocity graphs

    Activities: Biking to school, Library trip

Assign Project 1. E.g., Roller coaster.

Week 2

Top-down design (as in outline of an English paper, drawing a graph, i.e., when is it increasing, decreasing, concave up concave down?)

    Activities: A formula for piece-wise linear graphs or Given velocity graph, sketch distance graph

Rate graphs, Average rate of change

    Activity: Water tank (Other possibilities include: Slopes and difference quotients)

Weeks 3, 4

Definition of function

    Activity: Introduction to functions

    Word problems that involve setting up functions (top-down approach used)

    Postage function and idea of discontinuity

    Activity: Postage

More limits and continuity. One-sided limits. Average change versus marginal change. Definition of limit.

Properties of limits and continuity (algebraic and graphical)

Review of graphical calculus

    Translate graphs into words

    Activity: Dallas to Houston

    Given graph, draw the rate graph

    Activity: The leaky balloon

    Given a rate graph, draw the graph of quantity

    Activities: Given velocity graph, sketch distance graph (or The start-up firm)

    Concavity

Activity: Tax rates and concavity

Increasing, decreasing, concave up, concave down for functions represented as tables

Assign Project 2. E.g., The detector

Weeks 5, 6, 7

Slope of tangent as the limit of secant line slopes. (Refer back to the activity: Graphical estimation of slope.)

Definition of derivative and relation to slope

    Activity: Linear approximation. (Other possibilities include Slope with rulers.)

Derivative as rate of change

    Activity: Estimating cost

Function-derivative pairs

    $f \longleftrightarrow f'$

    distance $\longleftrightarrow$ velocity

    graph $\longleftrightarrow$ slope graph

    $q \longleftrightarrow$ rate of change of q

    cost $\longleftrightarrow$ marginal cost

Derivative of sine and cosine, including $\sin(k\,t)$

    Activity: Ferris wheel

Product, power, and quotient rules for differentiation

Assign Project 3. E.g., Water evaporation

Weeks 8, 9

Composition and the chain rule

    Activities: Graphical composition, Magnification, Using the product rule to get the chain rule

Representations of the derivative through formulas, graphs, tables

Higher order derivatives

    Activity: Finite differences

Intermediate value theorem

Activity: Can we fool Newton?

Weeks 10, 11

Using f' and f'' to graph f, find maxima, minima, and points of inflection

Activities: Using the derivative, Function derivative pairs

Optimization problems

Exponential and log functions

Activities: Why mathematicians use e^x, Exponential differences

Implicit functions, derivatives of rational powers, related rate problems

Assign Project 4. E.g., Rescuing a satellite.

Weeks 12, 13, 14

Antiderivatives and applications

Definition of definite integral and Riemann sums

Fundamental theorem of calculus

Activity: Fundamental theorem of calculus

Mean value theorem

Activity: Gotcha

Indefinite integrals

Differential equations

Activities: Animal growth rates, The product fund, Exchange rates

Calculus 2

Weeks 1,2

Review of definition of integral and Riemann sums

Activities: Time and speed, Oil flow

Geometric integrals

Activity: Graphical integration

Distance-velocity

> Activity: Can the car stop in time?

u–substitution

Integration by parts

Assign Project 1. E.g., Tax assessment.

Review of procedure to determine area between curves, including approximation by Riemann sums

> Activity: How big can an integral be?

Applications of integration (e.g., volume—at least one method chosen from cross-section, washer, shell)

Weeks 3, 4, 5

Differential equations

> Separable equations
>
> Exponential growth
>
>> Activities: Population, Fitting exponential curves, Log-log plots, Using scales
>
> Direction fields
>
>> Activities: Direction fields, Using direction fields, Drawing solution curves, The hot potato, Spread of a rumor. (If you do not assign "The fish pond" project, you might also use the activity "Save the perch" at this point.)
>
> Logistic equation
>
>> Partial fractions

Assign Project 2. E.g., The fish pond.

Weeks 6, 7, 8

More applications of integration. (E.g., arc length and/or surface area.)

Integrating from tables of values

Numerical integration

> Activity: Numerical integration

Trapezoidal rule, Simpson's rule

> Activity: Verifying the parabolic rule

More activities: Finding the average rate of inflation, Cellular phones, The shortest path, The River Sine

Assign Project 3 (end of week 7). E.g., Dome support in a sports stadium.

Week 9

Inverse functions

    Activity: Inverse functions from graphs

Derivatives of inverse functions

    Activity: Inverse functions and derivatives

Using inverse trig functions

Using tables of integrals

Week 10

Taylor polynomials

    Activities: Approximating functions with polynomials, Graphs of polynomial approximations,

L'Hôpital's rule

Improper integrals

    Activity: Convergence

Week 11

Sequences

    Activity: Sequences

Series as sequences of partial sums

    Activity: Investigating series

Tests for convergence: comparison test and integral test

    Harmonic series, p-series

Assign Project 4. E.g., Drug dosage.

Week 12

Geometric series and power series

> Activities: Comparing series and integrals, Space station, Introduction to power series. (Other possibilities include: Using series to find indeterminate limits, Using series to solve a differential equation.)

Ratio test

Algebra of convergent series

> Activity: Decimal of fortune

Weeks 13, 14

Taylor remainder, Taylor series, and applications

> Activities: Taylor series. (Other possibilities include: Using Taylor series to approximate integrals, Padé approximation, Second derivative test, Approximating logs.)

Complex numbers

> Activity: Complex power series.

Appendix B

Sample Questions

What follows is a complete set of quizzes and exams that have been used in our two-semester course. Quizzes were given in class and typically take 15 to 25 minutes. Exams were given in the evening, essentially untimed; final exams are two and one half hours long. Clearly, some of the questions look familiar, but many of the items are not "traditional." The reader may also note that some of the questions look like projects and activities. In fact, some were included to test the students' understanding of topics that had been introduced through projects and activities; in other cases, it was the other way around: an activity (or project) grew out of a quiz (or exam) question.

Following the quizzes and exams are a few more non-traditional questions.

Calculus 1, Quiz 1

1. The graph at the right gives the distance from Wichita of a car driving from Wichita to Kansas City. When time was 0 the car was in Wichita. The graph gives the velocity for five hours.

    (a) Where was the car two hours after it started?

    (b) What was the velocity of the car two hours after it started?

    (c) Write a brief paragraph that describes what the car did during the five hours.

    (d) Sketch the graph of the *velocity of the car* versus time for these five hours.

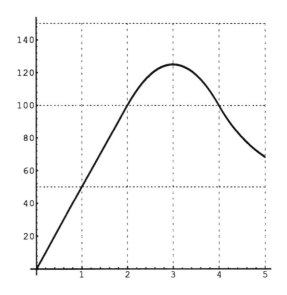

distance from Wichita (miles) vs. time (hours)

2. The cost of making three hats is $25, and the cost of making 12 hats is $70. Using cost as the vertical axis and number of hats as the horizontal axis the graph is *linear*.

    (a) Find the equation of the line.

    (b) Interpret what the slope means in terms of the problem.

Calculus 1, Quiz 2

1. A weather balloon is rising through the atmosphere. A thermometer on board reads 10° when the balloon is released from a ship on the ocean. The rate at which the temperature is changing is a *linear function* of the altitude (height above the ocean). The temperature was decreasing at a rate of 3° per kilometer when the balloon was one kilometer above the ocean and the temperature was decreasing at the rate of 5° per kilometer when the balloon was two kilometers above the ocean.

    (a) Sketch a graph of the rate of change of the temperature as a function of the altitude.

(b) Give a rule or formula for the rate of change of the temperature as a function of the altitude.

(c) Sketch a graph of the temperature as a function of the altitude.

(d) Give a rule or formula for the temperature as a function of the altitude.

2. Which of the following graphs could be the graph of $y = (x-3)^2$? Explain why.

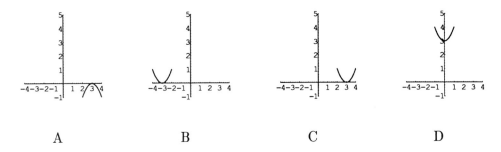

     A   B   C   D

Calculus 1, Quiz 3

A car passes a truck which is parked at a rest stop at 11:00 A.M. The car is driving along a straight section of a highway between Texas and New Mexico. The car is heading towards the New Mexico border, which is 500 miles from the rest stop. The velocity graph of the car from 11 A.M. until 4 P.M. is given below. At 12 noon the truck leaves the rest stop and heads towards New Mexico. The truck's velocity t hours after noon is $15t$ miles per hour.

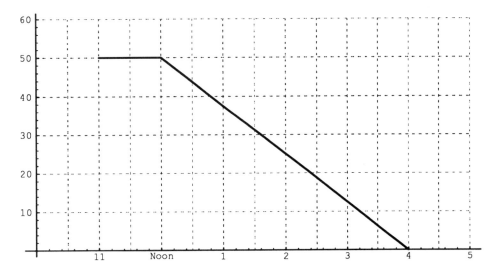

velocity (mph) vs. time of day

1. Sketch the graph of the *velocity of the truck* on the same set of axes as the velocity of the car.

2. Give a rule for the velocity of the *car* between 11 A.M. and 4 P.M.

3. At what time between 11 A.M. and 4 P.M. were the car and truck the furthest apart? Explain your answer. How far apart were they at that time?

4. Sketch a graph of the distance of the car from the rest stop versus time for time between 11 A.M. and 4 P.M.

5. (**Bonus**) Give a rule for the distance of the car from the rest stop as a function of time for time between 11 A.M. and 4 P.M.

Calculus 1, Quiz 4

The graph at the right gives the *rate* (in dollars per day) of change in value of a share of stock over the last two years. The stock was purchased on February 25, 1990, for $100 per share.

1. When was the stock worth the most during the last two years?

2. How much was the maximum value of the stock during the time period?

3. When was the stock worth the least during the last two years?

4. What was the least value of the stock during the time period?

5. Sketch the graph of the value of a share of stock during the time period shown.

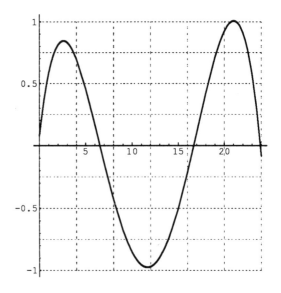

rate ($/day) vs. time (months)

Calculus 1, Quiz 5

1. (a) State the definition of the derivative.

   (b) Using only the definition of the derivative, compute the derivative of f at $x = 2$ for the function given by $f(x) = 3/x$.

2. If F is a function whose graph is given at the right, find the following if they exist. If the quantity does not exist, explain why.

   (a) $\lim\limits_{x \to 2} F(x)$

   (b) $\lim\limits_{x \to 4^-} F(x)$

   (c) $F'(3)$

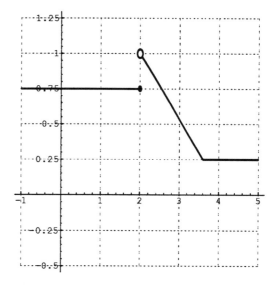

$y = F(x)$

3. If the profit that a company makes when it sells x dresses is x^3 dollars, find the *marginal profit* after the company has sold ten dresses. Explain what your answer means to the company.

Calculus 1, Exam 1

Please show all your work. You may use results from class but be sure to state the result. Be careful and read each question. Do not do unnecessary algebra. Be sure to label your graphs and give scales.

Note: In problems dealing with a graph, when you are asked for a numerical answer, a reasonable estimate is all that is required, but it is important that the geometric construction behind the answer be clearly illustrated on your graph. Effective labeling of the graph will often assist your explanations.

1. (10) The graph of a function f, which is defined for x in $[-2, 7]$, is given at the right. Use the graph to answer the following questions. Your answers should be of the form $a = \ldots$ For example, if you think f is discontinuous at $x = 2$, part of your answer to 1c would be $a = 2$.

    (a) Where (if anywhere) does $\lim_{x \to a^+} f(x)$ *not* exist?

    (b) Where (if anywhere) does $\lim_{x \to a} f(x)$ *not* exist?

    (c) Where (if anywhere) is f discontinuous?

    (d) Where (if anywhere) does the derivative of f *not* exist?

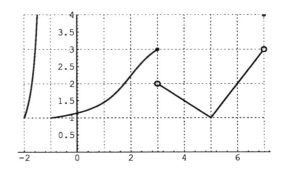

$y = f(x)$

2. (8) Compute the derivatives of the following functions:

    (a) $f(x) = (\sqrt{x})(\cos x)$
    (b) $g(y) = 3/\sin y$

3. (10)

    (a) Find the slope of the tangent line to the graph $y = 2x^5 - 12 \sin x$ at the point whose first coordinate is $x = 3$.

    (b) A car is driving along a straight highway from El Paso to San Antonio. The car starts at 12 noon in El Paso. The car has traveled $2t^5 - 12 \sin t$ miles by t hours after noon. How fast is the car traveling at 3 o'clock?

    (c) If a company sells x units of a product, its revenue will be $2x^5 - 12 \sin x$ dollars. What is the company's marginal revenue if it has produced 3 units of the product?

    (d) The volume that a gas occupies at a temperature of c degrees Celsius is $2c^5 - 12 \sin c$ cubic centimeters. What is the rate of change of the volume of the gas with respect to temperature when the temperature is 3° Celsius?

4. (15) Students start lining up outside the gym for registration at 8:00 A.M. at the rate of 200 students per hour. At 12 noon, the rate starts to gradually increase until 1 P.M. when the rate stabilizes at 300 per hour. At 1 P.M. when the gym opens, 1,050 students are in line. Once the gym opens, 250 students per hour are allowed to enter. After 2 P.M., the rate at which students arrive gradually decreases until 3 P.M. when the rate is 50 students per hour. After 3 P.M., students arrive at the rate of 50 students per hour.

   (a) Draw the graph of the *rate* at which students arrive versus time for time betweeen 8 A.M. and 4 P.M.

   (b) Draw the graph of the *rate* at which students are allowed to enter the gym versus time for time betweeen 8 A.M. and 4 P.M.

   (c) Draw the graph of the number of students who have arrived at the gym versus time for time betweeen 8 A.M. and 4 P.M.

   (d) Draw the graph of the number of students who have entered the gym versus time for time betweeen 8 A.M. and 7 P.M.

   (e) When was the line to enter the gym the longest? Explain.

   (f) What was the *average rate* at which students lined up during the time between 8 A.M. and 4 P.M.?

5. (12) A Ferris wheel has radius 30 feet, and when a car is at the lowest point on the wheel it is four feet above the ground. The wheel is revolving at 1 radian per second. Consider a car that is at the lowest point on the wheel at time 0.

   (a) Sketch a graph of the car's *height* above the ground for the next ten seconds.

   (b) How fast is the height of the car changing when $t = 3$ seconds?

   (c) At what time(s) is the height of the car changing the fastest?

Hint: Finding a rule for the height of the car above the ground may make 5b and 5c easier.

6. (20) The graph below gives the velocity of a car that is driving through Wichita to Kansas City. When time was 0, the car was in Wichita. The graph gives the velocity for five hours.

    (a) Write a brief paragraph that describes what the car did during the five hours.

    (b) Sketch the graph of the distance from Wichita versus time for these five hours.

    (c) At what time during the five hours was the car closest to Kansas City?

    (d) At what time between time = 1 and time = 5 hours was the car closest to Wichita? Explain.

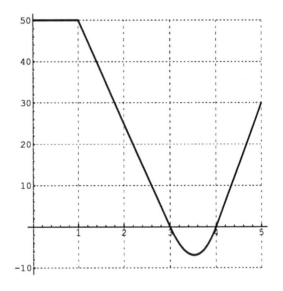

velocity (mph) vs. time (hours)

    (e) If the distance from Wichita to Kansas City is 150 miles, does the car reach Kansas City during the five hours? Explain.

    (f) Sketch the graph of the acceleration of the car versus time for these five hours.

7. (8) An economist is interested in how the price of a certain commodity (grain, for example) affects its sales. Suppose that at a price p per bushel, a quantity Q bushels of the commodity is sold. Let $Q = f(p)$.

    (a) Explain in economic terms the meaning of the statement $f(10) = 240,000$.

    (b) Explain in economic terms the meaning of the statement $f'(10) = -29,000$.

    (c) Use the above data to estimate the quantity sold if the selling price is $10.30. Be sure to explain your answer.

8. (9) The tables given below give values for three functions f, g, and h. All three functions are increasing on the interval $[1, 15]$.

x	$f(x)$	$g(x)$	$h(x)$
1	5	3	1
3	12	5	3
5	17	7	6
9	24	11	13
15	31	17	24

(a) Classify each function as bending up, bending down or increasing at a constant rate on $[1, 15]$. Be sure to explain your reasoning.

(b) Decide if the derivative of the function h has a largest value in $[1, 15]$. If the derivative does have a largest value, where does it occur? Explain.

9. (12) When a potato is heated in a conventional oven, the temperature of the potato after t hours in an oven set at 450° is given by the graph at the right. The potato's original temperature (before being put into the oven) was 70°.

(a) How long will it take for the potato to heat up to 300°?

(b) What is the rate of change of the temperature 45 minutes after the potato has been put in the oven?

If a potato is heated in a microwave oven, its temperature changes at the rate of 6° per minute.

(c) Sketch a graph of the temperature of a potato in a microwave oven. (Assume the temperature of the potato was 70° when it was put in the microwave.)

A potato is cooked when it reaches a temperature of 430°.

(d) Will one oven cook a potato faster than the other? Explain your answer.

(e) Is it possible to cook the potato faster by changing it from one oven to the other oven? Explain in detail.

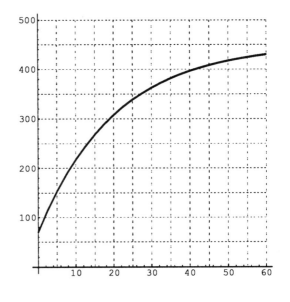

temperature (°C) vs. time (minutes)

Calculus 1, Quiz 6

The graph below gives the graph of the *acceleration* of a car for six hours. At time 0 the car was traveling west at 40 miles per hour and was 100 miles east of Chicago.

1. Give a rule for the *acceleration* for t in $[0, 6]$.

2. What was the *velocity* of the car at $t = 3$ hours?

3. When was the car's *velocity* the maximum? What was the maximum?

4. When was the car's *velocity* the minimum? What was the minimum?

5. Sketch the graph of the car's *velocity* for t in $[0, 6]$.

6. Sketch the graph of the *distance* of the car from Chicago for t in $[0, 6]$.

7. (**Bonus**) Give a rule for the *velocity* for t in $[0, 6]$.

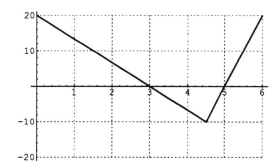

acceleration (mph per hour) vs. time (hours)

Calculus 1, Quiz 7

1. Compute the derivatives of the following functions:

   (a) $f(x) = (\sqrt{x})(\cos x)$

   (b) $g(y) = \sin y / \cos y$

2. A Ferris wheel has radius 40 feet, and when a car is at the lowest point on the wheel it is four feet above the ground. The wheel is revolving counter-clockwise at 1 radian per second.

   (a) Consider a car which is at the *highest* point on the wheel at time zero; sketch a graph of its *height* above the ground for the next six seconds.

   (b) How fast is the height of the car changing when $t = 3$ seconds?

   (c) At what time(s) is the height of the car changing the fastest?

Calculus 1, Quiz 8

1. (a) Compute the derivative of $\tan(\sqrt{x})$.
   (b) If the temperature of a chemical mixture is $\tan(\sqrt{x})$ degrees Celsius after x hours, what is the rate of change of temperature after $\pi/4$ hours?
   (c) What will happen to the chemical mixture of 1b as $x \to \pi/2$?

2. Let $f'(x) = (x-2)^2(x+4)^2$.
   (a) Where is the graph of f increasing? Where is the graph of f decreasing?
   (b) Locate any local maxima or minima of f.
   (c) Find any points of inflection of f.

Calculus 1, Quiz 9

1. Suppose the population of a county t years after 1980 is $500 + 50e^{0.2t}$.
   (a) At what rate is the population growing in 1985?
   (b) What was the population in 1970?

2. Let $f(x) = 3x^4 - 32x^3 + 96x^2$.
   (a) On what intervals is f increasing?
   (b) On what intervals is f decreasing?
   (c) On what intervals is f concave up?
   (d) On what intervals is f concave down?
   (e) List any local maxima, local minima, or points of inflection. (Indicate which is which.)
   (f) Sketch the graph of f.

Calculus 1, Quiz 10

1. Let $f(x) = x^2 \sin x + 5$.
   (a) Find the equation of the tangent line to the graph $y = f(x)$ at the point where $x = 2\pi$.
   (b) Use the tangent line you found in 1a to estimate the change in f if x increases from 2π to $2\pi + 0.01$.

2. The circumference of the earth at the equator is 25,000 miles. If a band were stretched around the earth at the equator it would be 25,000 miles long. If you wanted to lift this band so that it was one foot above the earth around the entire earth, about how much longer would the band have to be? Explain.

3. Let $F(x) = 16e^{0.05x}$.

    (a) What is the rate of change of F when $x = 10$?
    (b) What is the percentage rate of change of F when $x = 10$?
    (c) What value of x solves $F(x) = 14$?

Calculus 1, Quiz 11

1. A car is traveling at 60 feet per second when the driver spots a deer 300 feet ahead in the road and slams on the brakes. The following readings of the car's speed at various times are given:

Time	0	2	4	6	7
Speed	60	50	30	12	0

    where time is seconds since the driver slammed on the brakes and the speeds are in feet per second.

    (a) Give an estimate for the minimum distance the car will travel after the brakes are applied. Show your work. State any assumptions you use.
    (b) Give an estimate for the maximum distance the car will travel after the brakes are applied. Show your work.
    (c) If the deer freezes and does not move, will the car hit the deer? Explain.

2. If the car was traveling at 60 feet per second when it saw the deer and it slowed down at the rate of eight feet per second per second, would it hit the deer? Explain.

Calculus 1, Quiz 12

1. Find the area of the region bounded by $x = 2$, $x = 5$, the x-axis, and the curve $y = 3x^2 - 2\sqrt{x}$.

2. If the velocity of a car coasting along a straight road after t seconds is $50/t$ feet per second, how far will the car travel between 2 seconds and 10 seconds?

3. The temperature in an oven is changing at the rate of $10 \sec x \tan x + 0.01 e^x$ degrees per hour x hours after the oven is lit. How much will the temperature change during the first one and one half hours after it is lit?

4. If the marginal cost of making shirts at level of production x is $p(x)$ and $s(x)$ is an antiderivative of $p(x)$, then the total change in cost if production is increased from five shirts to 16 shirts will be (choose one):

(a) $p(16)$

(b) $p(16) - p(5)$

(c) $s(16)$

(d) $s(16) + C$

(e) $s(16) - s(5)$

Calculus 1, Quiz 13

1. (a) A farmer wishes to construct two adjacent rectangular fields, each of area 30,000 square feet, to be enclosed by fence as shown in the diagram. Find the dimensions of the fields which require the least amount of fencing.

    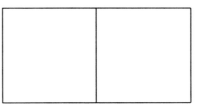

    (b) A farmer wishes to construct two adjacent rectangular fields, each of area A square feet to be enclosed by fence as shown in the diagram. Find the dimensions of the fields which require the least amount of fencing.

2. Decide whether the mean value theorem (MVT) is applicable to the functions on the given interval. If the MVT is applicable, find at least one of the points c.

    (a) $f(x) = 1/x$ on the interval $[2, 5]$

    (b) $f(x) = 1/x$ on the interval $[-1, 1]$

Calculus 1, Exam 2

Show all your work in your exam book.

1. (10) Evaluate the following integrals:

    (a) $\int (x^3 - 2x + \frac{4}{x} + 3)\,dx$

    (b) $\int (\cos x - 4\sqrt{x} + 12 \sec x \tan x + 4e^x)\,dx$

2. (10) Compute the derivatives of the following functions:

   (a) $f(x) = \tan(\sqrt{x})$
   (b) $g(t) = e^{4t}\cos(t^3)$

3. (7) Find the area enclosed by the x–axis, the vertical lines $x = 2$ and $x = 5$, and the curve $y = 2xe^x$.

4. (8) Find the equation of the tangent line to the graph of $y = (x^2 + 3x)^4$ at the point whose first coordinate is $x = 2$.

5. (10) Suppose the cost of making x machine tools is $100 + 3\ln(x^2 + 10)$ dollars.

   (a) What is the cost of making 100 machine tools?
   (b) What is the marginal cost if 100 machine tools have been produced?
   (c) What is the average cost of a machine tool if 100 machine tools are produced?
   (d) What happens to the marginal cost as $x \to \infty$? Do you think that the cost function is realistic? Explain.

6. (8) A train is traveling due west with velocity $v(t) = 60 + 10\sin t$ miles per hour after t hours, and it passes through Podunk two hours after it starts its trip.

   (a) How fast is the train going five hours after it started its trip?
   (b) How far is the train from Podunk five hours after the trip started?

7. (10) Define the function $G(x) = \int_1^x \dfrac{1}{t^2}\,dt$.

   (a) What is $G(1)$?
   (b) Draw a figure and indicate the area that would give the value of $G(2)$.
   (c) Use two rectangles and give an upper estimate and a lower estimate for the value of $G(2)$.
   (d) Give a formula for $G'(x)$.
   (e) Sketch the graph of G.

8. (8) The table below gives the *rate* at which oil is being imported by a country at selected times for the last three years for which data is available. (Amounts are in billions of barrels per year.)

Date	Rate
Jan 1, 1988	50
Apr 1, 1988	48
July 1, 1988	47
Oct 1, 1988	46
Jan 1, 1989	42
Apr 1, 1989	40
July 1, 1989	36
Oct 1, 1989	35
Jan 1, 1990	32

The energy minister has also stated that the rate at which oil was imported has been decreasing since January 1, 1988.

(a) Give the best *estimate* for the amount of oil imported in *1989*, and explain how accurate your estimate is.

(b) Let $T(x)$ be the *total amount* of oil that has been imported x years after January 1, 1988. Decide whether the graph of T is increasing and concave up, increasing and concave down, decreasing and concave up, or decreasing and concave down in the interval $[0, 2]$. Justify your choice.

9. (12) 9. Let $f(x) = \dfrac{3x^5 - 20x^3}{32}$.

   (a) Show that $f'(x) = \dfrac{15x^2(x-2)(x+2)}{32}$.

   (b) For the rest of this problem, you may use result 9a even if you were unable to obtain it on your own. You may also use the fact that $f''(x) = \dfrac{15x(x - \sqrt{2})(x + \sqrt{2})}{8}$.

   (c) On which intervals is f increasing and decreasing?

   (d) On which intervals is f concave up and concave down?

   (e) Classify each critical point of f as a local maximum, local minimum, or neither.

   (f) Where are the points of inflection of f?

   (g) Sketch the graph of f.

10. (12) *Consumer reports* is testing the braking performance of two cars: the Maxima and the Infiniti. (Notice how expensive cars are named after mathematical ideas.) The test consists of using the brakes to bring the car to a complete stop when the car is traveling at 60 miles per hour. The Maxima slows down at the uniform rate of 8 miles per hour per second. The Infiniti's *velocity* is given in the graph below.

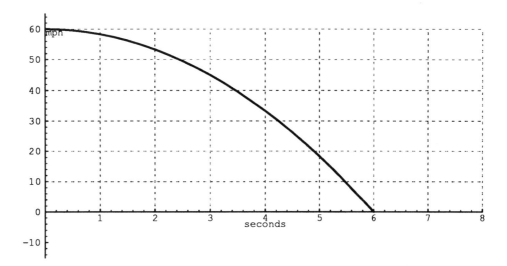

velocity (mph) vs. time (seconds)

(a) How much time will it take each car to stop?

(b) How far will each car travel before it stops?

(c) Write a brief paragraph summarizing braking performance which will be inserted into the report comparing these cars. If your data indicates that one of the cars is superior, explain your conclusion.

(d) (Optional) Pick a name for a new car model based on Calculus 1.

11. (10) You are involved in a research project that involves working with a species of laboratory animals. If $W(t)$ is the weight (in ounces) of such an animal t weeks after birth, then the growth of a healthy animal can be modeled by the differential equation $W'(t) = 10/W(t)$ (i.e., $\frac{dW}{dt} = 10/W$). You are responsible for an animal that weighs five ounces one week after it was born. So, according to the model the animal is growing at the rate of 2 ($= 10/5$) ounces per week when $t = 1$ week.

(a) Give the equation of the tangent line to the graph of $W(t)$ at the point $(1, 5)$.

(b) Use the tangent line to approximate the weight of the animal eight days after it was born.

(c) Classify the graph of $W(t)$ for $t \geq 1$ as increasing and concave up, increasing and concave down, decreasing and concave up, decreasing and concave down.

Calculus 1, Final Exam

Show all your work on these pages. Be sure to label your answers. To receive full credit for a problem, you must not only obtain the correct answer but also demonstrate that you can work similar problems correctly.

1. Find the derivatives of the following functions:

    (a) $f(x) = 3x^2 - \dfrac{2}{x^3} + \sqrt{x} - \pi^{\frac{2}{3}}$

    (b) $g(t) = 4t^2 \cos t$

    (c) $h(x) = \dfrac{2x}{1 + \ln(x)}$

    (d) $f(x) = e^{-3x+4}$

    (e) $f(x) = \sin^5(2x^3 - 1)$

2. Use the graph of the function f (below) to answer the following questions. If the quantity asked for does not exist, explain why.

    (a) $\lim\limits_{x \to 2} f(x)$

    (b) $\lim\limits_{x \to 4} f(x)$

    (c) $\lim\limits_{x \to 6} f(x)$

    (d) Is f continuous at $x = 4$? Explain.

    (e) Find the following values if they exist.

        i. $f'(1)$

        ii. $f'(4)$

        iii. $\displaystyle\int_{0.5}^{1.5} f(x)\,dx$

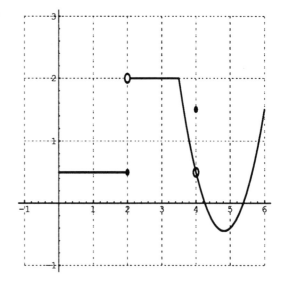

$y = f(x)$

3. Using the table of values for the function f (below), what is the best estimate for $f'(1.8)$? Explain your answer.

x	$f(x)$
1.0	3.0
1.2	3.9
1.4	3.7
1.6	4.2
1.8	4.3
2.0	4.4

4. A weather balloon is rising through the atmosphere. A thermometer on board reads 10° when the balloon is released from a ship on the ocean. The rate at which the temperature is changing is a *linear* function of the altitude (height above the ocean). The temperature is decreasing at a rate of 3° per kilometer when the balloon is 1 kilometer above the ocean, and the temperature is decreasing at the rate of 5° per kilometer when the balloon is 2 kilometers above the ocean.

   (a) Sketch a graph of the rate of change of the temperature as a function of the altitude.

   (b) Give a rule or formula for the rate of change of the temperature as a function of the altitude.

   (c) *Either* sketch a graph or give a formula for the temperature as a function of the altitude.

5. The graph below gives the velocity of a car traveling along a straight highway. The car is 300 miles away from Austin and traveling toward Austin at $t = 0$.

(a) Write a brief paragraph describing the location of the car during the eight hours shown.

(b) The speed limit is 60 miles per hour. Does the car ever travel at or exceed the speed limit? If so, for how long does the car travel at or exceed the speed limit?

(c) When is the car closest to Austin? Explain.

(d) When is the car furthest away from Austin? Explain.

(e) Does the car reach Austin during the eight hours? Explain.

(f) Sketch a graph of the location of the car for the eight hours.

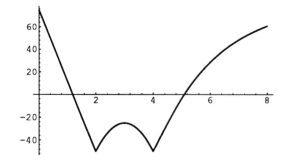

velocity (mph) vs. time (hours)

6. Sketch the graph of $f(x) = \dfrac{2x^2}{9-x^2}$ indicating local maxima and minima, inflection points, anymptotes, and finding the intervals where the function increases (and decreases) and where the function is concave up (and concave down). (Note that $f'(x) = \dfrac{36x}{(9-x^2)^2}$ and $f''(x) = \dfrac{108(x^2+3)}{(9-x^2)^3}$.)

7. The radius of a sphere is measured as 3 cm with a possible error of ± 0.5 cm. If the volume of the sphere is calculated using the formula $V = (4/3)\pi r^3$, use differentials to approximate the possible error in the computed volume.

8. Evaluate the following limits if they exist.

(a) $\lim\limits_{x \to 2} \dfrac{x^2 + x - 6}{x - 2}$

(b) $\lim\limits_{x \to 3} [x]$ ($[x]$ is the greatest integer function)

(c) $\lim\limits_{x \to \infty} \dfrac{\sin(x^2)}{x}$

9. Evaluate the following antiderivatives and definite integrals.

(a) $\displaystyle\int_{-1}^{2} (x^3 + \sqrt{x})\, dx$

(b) $\displaystyle\int (\cos(2x) + e^{-x})\, dx$

(c) $\displaystyle\int \dfrac{3x^2 - 4}{x^3 - 4x + 6}\, dx$

(d) $\displaystyle\int \sin(x)\cos(x)\, dx$

10. Assume that the rate of change of the population of sea birds in a bird sanctuary is proportional to the population of sea birds in the sanctuary. When the sanctuary was opened in 1961, there were 1000 sea birds there. In 1973 there were 2500 sea birds in the sanctuary. How many sea birds were in the sanctuary in 1990?

11. (a) Show that $d/dx(x \ln(x)) = 1 + \ln(x)$.

(b) Find $\int \ln x\, dx$. Show your work.

12. (a) Define the derivative of a function.

(b) Using only the definition of the derivative, compute the derivative of the function $f(x) = x^2 - 2x + 5$.

13. At the right is the graph of $f'(x)$.

(a) Are there any places where $f''(x)$ does not exist?

(b) Suppose $f(0) = 0$. Find $f(2)$.

(c) At what values of x does f have:

   i. local minima
   ii. local maxima
   iii. inflection points

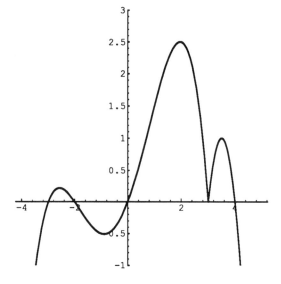

$y = f'(x)$

14. I am at point A on an east–west road that borders a forest. I have to reach a cabin at point B in the forest, 2 km east and 1 km north of A. If my jogging speed is 10 km/hr on the road and 6 km/hr in the forest, at what point P should I leave the road to reach the cabin in the minimum time?

Calculus 2, Quiz 1

1. Evaluate the following integrals.

   (a) $\int (3x^2 - 5)\sqrt{(x^3 - 5x + 6)}\,dx$

   (b) $\int_2^5 \dfrac{x + 2\,dx}{x^2 + 4x - 4}$

2. The graph below gives the *velocity* of a car driving west on a straight highway. Use the graph to answer the following questions.

(a) How far did the car travel between $t = 1$ and $t = 3$?

(b) How far did the car travel between $t = 3$ and $t = 5$?

(c) How far did the car travel between $t = 1$ and $t = 5$?

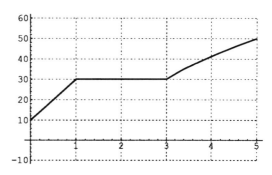

velocity (mph) vs. time (hours)

Calculus 2, Quiz 2

1. (a) Sketch the region bounded by the y–axis, $y = 9$, and $y = x^3 + 1$.
   (b) Compute the area of the region you found in 1a.
   (c) Express as an integral the *volume* of the solid formed by revolving the the region of 1a around the x-axis. *Do not evaluate the integral.*
   (d) Compute the volume of the solid formed by revolving the region of 1a around the y-axis.

Calculus 2, Quiz 3

1. (a) Sketch direction fields for the differential equation $dy/dt = y(y - 40)(y + 20)$.
   (b) List any constant solutions to the DE and classify each as an attractor or repellor.
   (c) Sketch the solution to the DE that passes through $y = 5$ when $t = 1$.
   (d) Sketch the solution to the DE that passes through $y = -5$ when $t = 1$.
   (e) Sketch the solution to the DE that passes through $y = 50$ when $t = 1$.

2. Assume that the rate of change of the temperature of a substance is proportional to the reciprocal of the temperature.

   (a) Express this situation as a differential equation for the temperature of the substance as a function of time.

(b) Find the general solution to the differential equation you found in 2a.

Calculus 2, Quiz 4

1. (a) Solve the differential equation $dy/dt = 2y - 3$, $y(0) = 1$.
   (b) Sketch a graph of your solution to 1a.

2. On an island, 3% of the population die each year and the birth rate is 5% per year. Each year 500 people move onto the island, but 700 people leave each year. The current population is 20,000.

   (a) Write a differential equation for P, the population of the island after t years.
   (b) What will happen to the population in the long run? Explain.

Calculus 2, Exam 1

Show all your work.

1. (20) Let R be the region between the curve $y = \sqrt{x-3}$, the x-axis, and the lines $x = 3$ and $x = 19$.

   (a) *Sketch* the region R.
   (b) Find the *area* of R.
   (c) Find the *volume* of the solid formed by revolving R around the x-axis.
   (d) Express the *volume* of the solid formed by revolving R around the y-axis as a definite integral. *Do not evaluate the integral.*

2. (10) Evaluate the following integrals:

   (a) $\int_0^1 \sec^2 x \tan^2 x \, dx$
   (b) $\int x e^{-x^2} \, dx$

3. (10) The graph below gives the annual rate of change (in percent per year) in the value of a baseball trading card from the beginning of 1981 to the end of 1986. The card sold for \$100 at the beginning of 1981.

(a) What was the value of the card at the end of 1981?

(b) What was the value of the card at the end of 1982?

(c) What was the average rate per year for the entire period?

(d) How much was the card worth at the end of 1986?

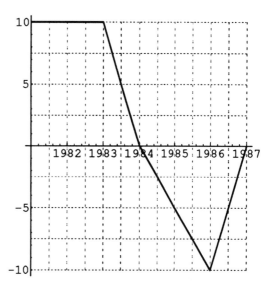

rate (% per year) vs. year

4. (10) Consider the differential equation $dy/dt = \ln(y)$.

   (a) Determine whether there are any steady state solutions.

   (b) Sketch the solution that passes through $y = 10$ when $t = 0$.

   (c) Sketch the solution that passes through $y = 1$ when $t = 0$.

   (d) Sketch the solution that passes through $y = 0.5$ when $t = 0$.

   For parts 4b, 4c, and 4d, indicate how you obtained your sketch.

   (e) What happens to the solutions in 4b, 4c, and 4d as $t \to \infty$?

5. (10) A teacher invested $5,000 today in a savings certificate that pays interest at 8% compounded continuously. She also has a bank account that pays 4% compounded continuously which has $4,000 in it today. She decides to deposit money in the bank account at the rate of $500 per year.

   (a) How much will the savings certificate be worth in five years?

   (b) How much will be in the bank account in five years?

   (c) When (if ever) will the bank account be worth more than the savings certificate?

   In parts 5b and 5c you should assume that no money is ever taken out of the bank account.

6. (10) The population of deer in a national park is modeled by the differential equation $dP/dt = kP^4$, where $P(t)$ is the population of the deer t years after 1990. In 1990 there were 100 deer in the park, and in 1993 there were 200 deer in the park.

    (a) When will there be 400 deer in the park?
    (b) How many deer will be in the park in the year 2,000? Explain your answer.

7. (10) The Wildlife Service is putting up a temporary fence in a river to keep large fish from eating newly stocked trout. The fence will be put in a part of the river that is 70 feet wide. The fence will be made of wire mesh which costs $2.50 per square foot. The following table gives the depth (in feet) of the river measured at 10-foot intervals at the location where the fence will be built. (In the table, x represents the distance from one bank of the river.) Estimate the cost of the wire mesh needed for the fence. Explain what types of river bottoms should give good results and what kind would give bad results.

x	0	10	20	30	40	50	60	70
Depth	2	21	30	42	30	36	22	8

    (**Bonus**) The bottom of the mesh (the part on the bottom of the river) must be lined with plastic that costs $1.25 per foot. Estimate how much plastic will be required.

8. (8) True or false? Explain your reasons for your choice.

    (a) If a region R is revolved around the x–axis, the volume of the resulting solid will be equal to the volume of the solid obtained by revolving R around the y–axis.
    (b) If dp/dt is *positive* for all t, then there must be values of t for which $p(t) > 1,000,000,000$.

9. You are the head of the integration department at a scientific company. One of your workers brings you a new procedure for estimating integrals. For your test of this procedure you use as a test problem the integral of $1,000x^4 \, dx$ from 1 to 3.

    (a) (2) Show that the correct answer to the test problem is 48,400.

    When you try out the new procedure for three different values of h you get the following results:

h	Estimate for $\int 1,000x^4 \, dx$	Error
1.00	48,666.6666	266.6666
0.50	48,416.6666	
0.25	48,401.0416	

(b) (2) So the exact error when h is 1 is 266.6666. Compute the exact error the new procedure made for the remaining values of h and fill in the table.

From your table above it is obvious that the error decreases as $h \to 0$. However, to compare the new procedure to any other procedures you need to know how the error depends on h. If you plot error on the vertical axis and h along the horizontal, it will be difficult to find a relation between error and h (try it if you like), so we will change the scale of the graph to see what is happening.

(c) (3) Fill in a table that corresponds to the data in your previous table:

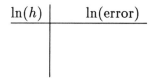

(d) (3) Plot $\ln(\text{error})$ along the vertical axis and $\ln(h)$ along the horizontal axis.

(e) (3) Find $\ln(\text{error})$ in terms of $\ln(h)$. That is, find a rule for $\ln(\text{error})$ as a function of $\ln(h)$.

(f) (3) Now find error in terms of h. (I.e., find the rule for error as a function of h.)

(g) (3) Discuss whether the new procedure might be better than the trapezoidal rule.

Calculus 2, Quiz 5

1. A car is traveling at 60 feet per second when the driver spots a deer 300 feet ahead in the road and slams on the brakes. The following readings of the car's speed at various times are given:

Time	0	2	4	6	7
Speed	60	50	30	12	0

   where time is seconds since the driver slammed on the brakes and the speeds are in feet per second.

   (a) Give an estimate for the maximum distance the car will travel after the brakes are applied. Show your work. State any assumptions you use.

   (b) If the deer freezes and does not move, will the car hit the deer? Explain.

2. The speed of a snail after t hours is $\sqrt{5t+1}$ miles per hour.

(a) Express the distance the snail travels as t goes from 0 to 2 hours as an integral.

(b) Use four intervals and the trapezoidal rule to approximate the integral.

(c) Use four intervals and Simpson's rule to approximate the integral.

Calculus 2, Quiz 6

1. Write the Taylor polynomials of degree 0, 1, and 2 for the function $f(x) = e^{-x^2}$ around $x = 0$.

2. (a) Write the Taylor polynomial of order 3 for x^3 around $x = 2$.

   (b) How well do you think the polynomial you found in 2a approximates the original function x^3. Explain.

Calculus 2, Quiz 7

1. Give a power series for the following functions around 0. (Give at least the first four non-zero terms, and state the values of x for which the series converges.)

   (a) $x^2 e^{-x}$

   (b) $\cos(\sqrt{x})$

   (c) $\frac{3}{1+3x^2}$

2. The budget this year for the Super Slider project is $5 billion. As part of the drive to shrink the national debt, the government has decided to cut the budget of this project by 10% each year. (This means that the budget for one year is cut by 10% to get the budget for the next year.)

   (a) What will the budget for the Super Slider be in 20 years?

   (b) What will the *total* budget expenses for the Super Slider over the next 20 years?

   (c) Is there an upper bound for the total budget expenses of the Super Slider? Explain.

Calculus 2, Quiz 8

For each of the following integrals or series decide whether the integral or series converges or diverges. If the integral or series converges, state the value of the limit if you can. Show all your work.

1. $\int_3^\infty \dfrac{5\,dx}{\sqrt{x}}$

2. $\int_3^\infty \dfrac{5\,dx}{x^3-1}$

3. $\sum_{n=1}^\infty \dfrac{1}{n^2}$

4. $\sum_{n=1}^\infty e^{-n}$

Calculus 2, Quiz 9

1. Consider the power series $3 + \sum_1^\infty (2n)(3x-4)^n$.

    (a) What is the radius of convergence of the power series?

    (b) For what values of x does the power series converge?

2. Let $f(x)$ be given by the graph at the right.

    (a) Decide whether f has an inverse.

    (b) Give the domain and range of the inverse if the inverse exists.

    (c) Find $f^{-1}(2)$.

    (d) Estimate $f^{-1\,\prime}(2)$.

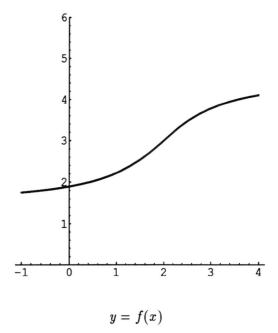

$y = f(x)$

3. Let $f(x) = e^{x^2}$ for $x \geq 0$. Let g be the inverse of f.

    (a) What is $g(e)$?

    (b) What is $g'(e)$?

    (c) What is $g'(1)$?

Calculus 2, Quiz 10

1. Compute the derivative of the function $f(x) = \sin^{-1}(4x^2)$.

2. Evaluate the following integrals.

   (a) $\displaystyle\int_{0.5}^{2} \frac{dx}{1+x^2}$

   (b) $\displaystyle\int_{0}^{0.5} \frac{2x\,dx}{\sqrt{1-x^2}}$

3. Construct a right triangle with hypotenuse x and one leg with length 1.

   (a) Label $\sec^{-1}(x)$ on the triangle.

   (b) Use the triangle to express the following in terms of x:

   i. $\cos(\sec^{-1}(x))$

   ii. $\tan(\sec^{-1}(x))$

Calculus 2, Exam 2

Part 1 (15)

As soon as you are done with this part, hand it in and pick up Part 2 and integral tables. Show all your work on this page.

1. $\displaystyle\int_{3}^{\infty} \frac{dx}{1+x^2}$

2. $\displaystyle\int x^2 e^x\,dx$

3. $\displaystyle\int \frac{5\,dx}{\sqrt{9-x^2}}$

Part 2

Show all your work in your bluebook. If you use the table of integrals give the number of the formula you are using.

1. (12) Consider the curve $y = 1/x$.

   (a) Sketch the region above the x–axis, under the curve and to the right of $x = 1$.

   (b) Decide if the area of this region is finite. If so, what is the area?

(c) Revolve the region around the x-axis. Is the volume of the solid finite? If so, what is the volume?

2. (10) Suppose the government pumps an extra $1 billion into the economy. Assume that each business and individual saves 25% of its income and spends the rest, so that of the initial $1 billion 75% is respent by individuals and businesses. Of that amount, 75% is respent, and so forth. What is the total increase in spending due to the government action?

3. (20) Consider the region below the curve $y = x^2 - 3$ and above $y = 1$.

    (a) Sketch the region described in 3.
    (b) Find the area of the region described in 3.
    (c) If the region in 3 is revolved around the x-axis, express the volume of the region formed as an integral. *Do not evaluate the integral.*
    (d) A solid is formed by putting isoceles right triangles on top of the region described in 3. A leg of each triangle is parallel to the y-axis, and each triangle is perpendicular to the x-y plane. Express the volume of the solid as an integral. *Do not evaluate the integral.*

4. (20) Write the first four non-zero terms of the Taylor series (power series) around 0 for the following functions. If there are any restrictions on your series (i.e., if the series is only valid for certain values of x), write these restrictions.

    (a) $f(x) = \cos x$
    (b) $f(x) = 1/(1-x)$
    (c) $f(x) = 1/(1+x^2)$
    (d) $f(x) = \tan^{-1}(x)$
    (e) $f(x) = e^{\sqrt{x}}$

5. (6) True or false? If you answer false, give a counterexample for full credit.

    (a) If $\lim_{x \to \infty} f(x) = 0$, then $\int_1^\infty f(x)\, dx$ converges.
    (b) If $f(x) > 0$ for all x, then there is an inverse function for f.

6. (6) Which of the following polynomials is the best approximation to the function $f(x) = e^x$ around $x = 0$? Explain your choice.

    (a) $1 + x + x^2$
    (b) $1 + x$
    (c) $1 + x + x^2/2$
    (d) $1 + x + x^2 + x^3$

7. (13) A forester has found that the height of a tree is a function of the age of the tree. (I.e., $h = f(a)$, where h is measured in feet and a is measured in years.) She has the following data:

age	height = $f(a)$	$f'(a)$
2	4.0	1.2
3	5.1	1.0
4	6.1	0.9
5	7.0	0.6
7	8.0	0.2

(a) When the tree is seven years old, how fast is it growing?

(b) When the tree is seven feet tall, how old is the tree?

(c) When the tree is seven feet tall, about how much time does it take for it to grow one *inch* taller? Explain.

Calculus 2, Final Exam

Answer all questions in your exam book. Show all your work. Include enough explanation of your work to show that you could also solve similar problems.

1. Evaluate each of the following integrals.

    (a) $\int e^{5x}\sqrt{1 + 6e^{5x}}\, dx$

    (b) $\int x^2 e^{-2x}\, dx$

    (c) $\int \dfrac{dx}{x\sqrt{9 + x^2}}$

    (d) $\int_0^\infty \dfrac{dx}{1 + x^2}\, dx$

2. (a) Find the area of the region enclosed by the curves $y = 6x - x^2$ and $y = 3x$.

    (b) Express as an integral, but do not evaluate, the volume of the solid generated when the region in 2a is revolved about the x–axis.

    (c) Express as an integral, but do not evaluate, the volume of the solid generated when the region in 2a is revolved about the line $x = -1$.

3. Solve $dy/dt = y^2 t^2$, where $y = 6$ when $t = 0$.

4. A population is growing according to the differential equation

$$\frac{dP}{dt} = 0.05\sqrt{10{,}000 - P^2}, \qquad P(0) = 5.$$

   (a) Solve the differential equation.

   (b) Does the population ever exceed 400? Explain.

5. Dead leaves accumulate on the forest floor at the rate of three grams per square centimeter per year, while they decay at a continuous rate of 75% per year. Write a differential equation that corresponds to this situation.

6. Suppose $N(t)$ is the number of people who remember a newly advertised product t days after viewing an ad. The rate of change of N is proportional to N, and 100 people see an ad for a new product. Three days later, only 40 of them remember the product. How long after the original ad do only 20 of them remember the ad?

7. (a) Find a power series expansion of $\sin\sqrt{x}$ about 0.

   (b) For what values of x will your series converge?

   (c) Use your series to evaluate $\displaystyle\int_0^2 \sin\sqrt{x}\,dx$.

8. At the right is the graph of f, whose power series around $x = 0$ is $f(x) = -1 + x/3 + x^2/2 - x^3/6 + x^4/12 - x^5/20 + \cdots$.

   (a) Sketch a graph of f^{-1}.

   (b) Find $f^{-1}(-1)$.

   (c) Find the derivative of f^{-1} evaluated at -1.

   (d) Give the Taylor polynomial of order 1 for f around -1.

   (e) (**Bonus**) Find the Taylor polynomial of order 2 for f^{-1} around -1.

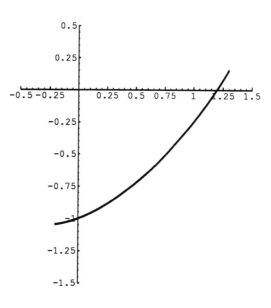

$y = f(x)$

9. If the Taylor series for $f(x)$ around $x = 2$ is
$$f(x) = 4 - 6(x-2)^2 + 12(x-2)^3 - 17(x-2)^5 + \cdots$$
which of the graphs below gives the best approximation fo the graph of f near $x = 2$. Explain your choice.

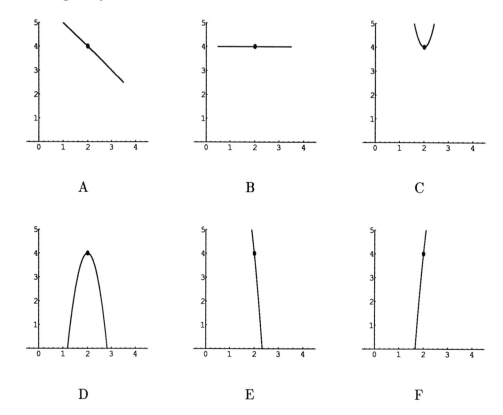

10. The six graphs below represent possible solutions to some differential equations. For each differential equation, (i) choose the graph that best represents a solution to the differential equation and (ii) give the equation of the asymptote of the graph you choose.

   (a) Answer (i) and (ii) for
   $$dy/dt = 0.05\, y\,(500 - y), \qquad y = 100 \text{ when } t = 0.$$
   Explain your choice.

   (b) Answer (i) and (ii) for
   $$dy/dt = -2y + 500 \qquad y = 100 \text{ when } t = 0.$$
   Explain your choice.

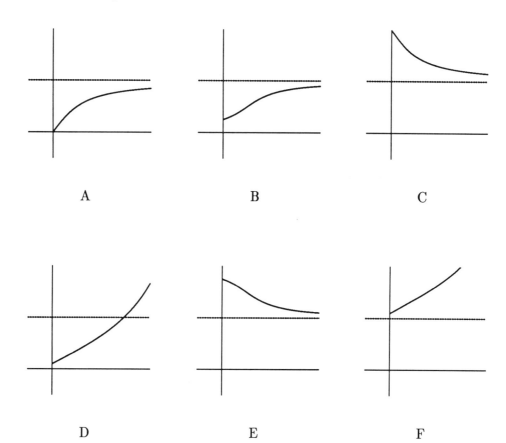

11. Which of the following series converge? For those that converge, determine the sum if you can.

(a) $\sum_{n=1}^{\infty} \dfrac{3}{n+2}$

(b) $\sum_{n=1}^{\infty} \dfrac{1}{n^2+1}$

(c) $\sum_{n=0}^{\infty} \dfrac{2^n}{n!}$

(d) $\sum_{n=0}^{\infty} \dfrac{3}{4^n}$

12. You own a plot of riverfront property which is pictured in the figure. Your property runs along the x-axis from $x = 0$ to $x = 100$ and is bounded by the lines $x = 0$, $x = 100$, and the River Sine whose equation is $y = 60 + 10\sin(x/5)$.

    (a) What is the area of the plot?

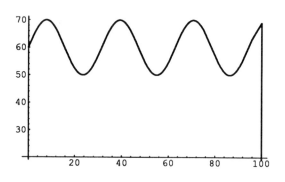

Your property

   (b) You have \$600 to spend, and you hire a gardening firm to install a fence along the riverside part of your property. The firm will charge you \$3 a foot for the fence including installation. If there is any money left over, the firm will fertilize the lot. They charge \$20 a bag for fertilizer. (This charge includes spreading the fertilizer.) One bag of fertilizer will cover 1,000 square feet.

      i. Will there be any money left for fertilizer after the fence along the river is installed? Explain your answer.

      ii. If there is money for fertilizer, about how much of your plot will be fertilized? Explain.

Miscellaneous Questions

1. Below is a graph (with some information missing) of a function $y = f(x)$. Suppose that in addition to the information we have on the graph we also know that the domain of the function is $[1, 4]$ and that the function has a second derivative for all x in the interval $(1, 4)$. Answer each of the following questions True, Impossible, or Maybe (i.e., we need more information to make an absolute determination).

(a) There is at least one number c in the interval $(1,4)$ such that $f'(c) = 0$.

(b) There is at least one number x_0 in the interval $(1,4)$ such that $f(x_0) = 0$.

(c) There is at least one number d in the interval $(1,4)$ such that $f''(d) < 0$.

(d) There is at least one number p in the interval $(1,4)$ such that $f''(p) > 0$.

(e) $f(1) = 2$.

(f) $f(2) = 1$.

(g) $\lim_{x \to 3} f(x) = 1$.

(h) There is at least one number q in the interval $(1,4)$ such that $f'(q) = -2$.

(i) There are two numbers a and b in the interval $(1,4)$ such that $\int_a^b f(x)\,dx = 0$.

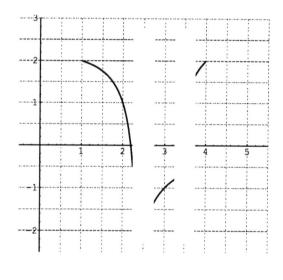

$y = f(x)$

2. Suppose $f(t)$ is the value of an automobile, t years from now.

   (a) Would you expect $f'(t)$ to be positive or negative? Explain in a sentence or two.

   (b) In ten years, would you expect $f''(t)$ to be positive or negative? Explain in a sentence or two. (Note: a sketch of the graph of f may help you.)

3. Show that $2 < \int_0^4 \dfrac{1}{1 + \sin 2x}\,dx < 4$. (Hint: Sketch the function.)

4. Consider the equation $\ln x = e^{-x}$.

(a) Explain how you know that there is some $c \in (0, 100)$ that satisfies the equation.

(b) Estimate c (within 1). Show enough work so that it is clear what you are doing.

(c) How do you know that there is only one c that satisfies the equation?

5. In watching you prepare for your calculus final exam, your roommate has drawn some conclusions about calculus. For each, explain why it is or isn't true.

   (a) $\int_0^1 f(x)g(x)\,dx = \int_0^1 f(x)\,dx \cdot \int_0^1 g(x)\,dx$ for all f and g.

   (b) $\sin 2x + 1$ and $\pi - \cos 2x$ can not both be antiderivatives of $2\sin x \cos x$ since they are not the same function.

   (c) If $\int_0^3 (f(x) - g(x))\,dx = 0$, then $f(x) = g(x)$ for all $0 \leq x \leq 3$.

6. Given the information in the table, compute the following derivatives.

   (a) $(f+g)'(4)$

   (b) $(f/g)'(3)$

   (c) $(f \circ g)'(2)$

	1	2	3	4
f	2	4	1	3
g	3	1	2	4
f'	4	2	3	1
g'	1	3	4	2

7. A class of calculus students was given the following table of values for a function f.

x	1	2	3	4	5	6
$f(x)$	4.2	4.1	4.2	4.5	5.0	5.7

   The students were asked to estimate the value of the derivative of f at the point 4. Three students gave the following as their approximations.

   Student 1 $f'(4)$ is approximately $\frac{f(5)-f(4)}{5-4}$.

   Student 2 $f'(4)$ is approximately $\frac{f(4)-f(3)}{4-3}$.

   Student 3 $f'(4)$ is approximately $\frac{f(5)-f(3)}{5-3}$.

   (a) What estimate did student 1 get?
   (b) What estimate did student 2 get?
   (c) What estimate did student 3 get?
   (d) Using the table above draw a rough sketch of the graph of f.

(e) How can the three different estimates be represented on this graph? Try to be as precise as you can.

(f) Which estimate do you think is likely to be best? Explain.

8. Consider the function whose graph is below.

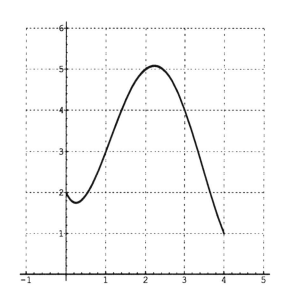

Estimate $\int_0^4 f(x)\,dx$ using

(a) R_4 (i.e., right-hand rule with $n = 4$)

(b) T_2 (i.e., trapezoidal rule with $n = 2$)

(c) S_2 (i.e., Simpson's rule with $n = 2$)

$y = f(x)$

9. I used my calculator to estimate some integral and obtained the following *errors*. I can't remember whether I used the left-hand rule, midpoint rule, or Simpson's rule.

n	Error
5	0.197397
10	0.038443
50	0.001404
100	0.000350

(a) Which method do you think I used? Explain.

(b) What n would give an error less than 0.000001? Explain.

10. The table below gives values of $f(x)$ for various values between 0 and 1.

x	$f(x)$	x	$f(x)$
0.000	2.000	0.625	0.916
0.125	1.938	0.750	0.649
0.250	1.765	0.875	0.433
0.375	1.510	1.000	0.271
0.500	1.213		

(a) Use Simpson's method with $n = 4$ to approximate $\int_0^1 f(x)\,dx$.

(b) If you were to approximate the integral using $n = 20$, by how much would you expect the error to shrink.

11. Before Galileo discovered that the speed of a falling body (with no air resistance) is proportional to the time since it was dropped, he mistakenly conjectured that the speed was proportional to the distance it had fallen.

    (a) Assume the mistaken conjecture to be true and write an equation relating the distance fallen $D(t)$, at time t, and its derivative.

    (b) Using your answer to 11a and the correct initial condition, show that D would have to be equal to 0 for all t, and therefore the conjecture is wrong.

12. (a) Verify that the first two non-zero terms in the expansion
$$\tan x = x + x^3/3 + 2x^5/15 + 17x^7/315 + \ldots$$
are correct.

    (b) Find the first four non-zero terms in a power series expansion for $sec^2 x$.

    (c) Find the first four non-zero terms in a power series expansion for $\ln|\sec x|$.

13. The graph below gives the velocity (in miles per hour) of a vehicle driving along a straight road from point A toward B (which is 500 miles from A) for six hours. When time $t = 0$, the vehicle is 100 miles away from A and 400 miles away from B. (A negative velocity means the vehicle is traveling towards point A and away from point B.) The kinetic energy of the vehicle is $0.01v^2$ where v is the velocity of the vehicle.

    (a) Sketch the graph of kinetic energy versus time for the six hours. (Be sure to label the axes of your graph.)

    (b) What is the rate of change of kinetic energy with respect to time when $t = 1.5$?

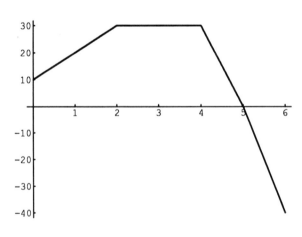

velocity (mph) vs. time (hours)

14. Use the graph of f' below to answer the following questions. You are also given that $f(-2) = 1$. (Note: The graph given is the graph of f', the derivative of f.)

   (a) True or false:

   i. $f(-2) = 0$. (False, since $f(-2) = 1$.)

   ii. $f'(-2) = 0$. (True, since $(-2, 0)$ is a point on the graph of f'.)

   iii. The function f has a local minimum at $x = -2$.

   iv. The graph of f is concave down for $x > 1$.

   v. The function f is increasing for $-2 < x < 2$.

   vi. The graph of f has an inflection point at $x = 1$.

   vii. The function f has a critical point at $x = -1$.

   (b) Sketch the graph of f''.

   (c) Sketch the graph of f.

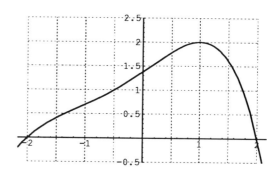

$y = f'(x)$

15. At the right is a sketch of the function f. Approximate $\int_{2}^{7} f(x)\,dx$ with a Riemann sum using the partition $P: 2, 4, 6, 7$.
Pick $\overline{x_i}$ so that $f(\overline{x_i})$ is the maximum of the function f over the ith interval. (I.e., find an upper Riemann sum.)

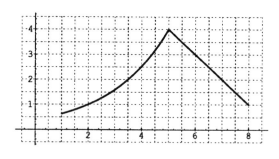

$y = f(x)$

16. A camcorder is on level ground 1000 feet from where a rocket is being launched vertically. The camcorder is set up to automatically track the rocket. The graph below gives the angle of inclination of the camcorder, θ, as a function of the time elapsed since the rocket was launched.

    (a) What is the value of θ after five seconds?

    (b) At what rate is the angle of inclination of the camcorder changing after 5 seconds?

    (c) At what rate is the distance of the rocket from the ground changing after 5 seconds?

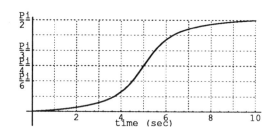

$y = \theta(t)$

17. For each of the following, the graph of the function f is sketched. On the same set of axes, sketch the graph of the derivative, f', where it exists.

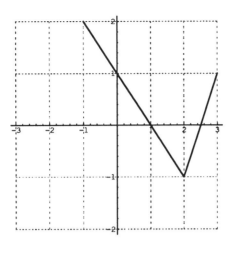
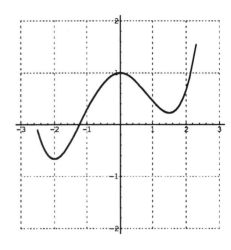

A   B

18. According to the December 2, 1990, *New York Times*, "For a few years after the stock market's plunge in September, 1987, employment in New York City held steady at around 3.6 million. Then it fell. And now it is falling faster."

    Draw a graph that illustrates the information given above. Be sure to label the axes of your graph.

19. This problem concerns $\lim_{h \to 0} \dfrac{(4+h)^{1/2} - 2}{h}$.

    (a) The above limit is actually the definition of $f'(c)$ for what function f at what number c?

    (b) Using any method you wish, find the value of this limit.

20. People start lining up for tickets to a play at 5 P.M. They line up at the rate of 50 per hour until 6:30; then the rate at which they line up increases for the next 30 minutes until at 7 P.M. there are 120 people in line. After 7 P.M. the rate at which people arrive at the ticket window is 70 per hour until 8 P.M., and then it gradually decreases to 0 per hour by 9 P.M. At 7:30 the ticket window opens and starts selling tickets at the rate of 60 per hour.

    (a) What time should you get in line to have the shortest wait?

    (b) When was the line for tickets the longest?

    (c) (Bonus) There were 300 tickets left for the performance when the ticket office opened. Were all the tickets for the performance sold? Explain.

21. The table below gives the rate of change of temperature with respect to altitude. The temperature at an altitude of 30 km is 250° K.

Altitude	30	32	34	36	38	40
Rate of change of temperature	5.0	4.4	4.0	3.8	3.4	2.5

All altitudes are in kilometers and the rates are in degrees K per kilometer. You may assume that the rate of change of temperature is decreasing between 30 and 40 kilometers. We want to find the temperature at an altitude of 40 km.

(a) What is a minimum for the temperature at 40 km? Explain.

(b) What is a maximum for the temperature at 40 km? Explain.

(c) What is your best guess for the temperature at 40 km? Why is it the best?

(d) Is your answer to 21c the actual temperature at 40 km? Explain.

22. The ideal gas law says that $pV = cT$, where c is a constant, p is pressure, V is volume, and T is the temperature of a fixed amount of gas. Find the rate at which the volume is changing with respect to pressure when $p = 12$ (atmospheres) and $V = 16$ (liters) if the temperature is kept fixed.

23. The price of a barrel of oil is currently $30. The price is predicted to change at the rate of $10 \sin t$ dollars per week t weeks from now for the next ten weeks. Assuming this prediction is correct:

(a) Give a rule for the price of a barrel of oil in terms of t, the number of weeks from now.

(b) Sketch a graph of the price of a barrel of oil versus time for the next ten weeks.

(c) Will the price of a barrel of oil go above $43 during the next ten weeks? Explain.

24. Use your calculator to calculate values and organize the values into a table. Use the table to find the value of $\lim_{x \to 0} \frac{\tan(\sqrt{x})}{x}$. If the limit does not exist, say so.

25. An interplanetary probe is sending back information about temperatures on a newly discovered planet. The probe sends back two different types of information: (1) readings on the temperature and (2) readings on the rate of change of the temperature with respect to altitude. Because of electrical interference, not all the data is received. The data received has been displayed on the two graphs given below. Answer the following questions and explain your work.

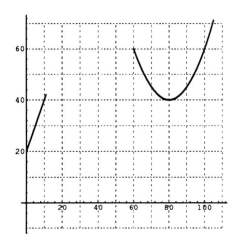
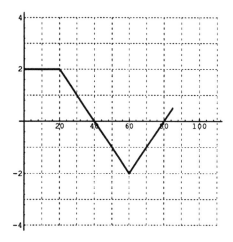

temperature (°C) vs. altitude (km)   rate of change of temperature (°C/km) vs. altitude (km)

(a) What is the temperature when the altitude is 10 km?

(b) At what rate is the temperature changing when the altitude is 10 km?

(c) What is the temperature when the altitude is 40 km?

(d) At what rate is the temperature changing when the altitude is 100 km?

(e) At what altitude is the temperature minimized? (I.e., at what altitude is it the coldest?)

(f) Fill in as much as possible of the missing parts of *both* graphs.

26. (a) On the grid below, draw the graph of $y = 20\sin(2t)$ for t in $[0, 10]$. Be sure to label the axes.

(b) Estimate the slope of the graph when $t = 1$.

(c) If your velocity t seconds after you start walking is given by $v(t) = 20\sin(2t)$ feet per second, estimate how far you travel in the time interval between $t = 1$ second and $t = 3$ seconds.

27. A weather balloon is released from a weather station, and as it rises through the atmosphere it radios back temperature readings. The readings are graphed at the right.

    (a) How fast is the temperature changing when the altitude is 3 km?

    (b) At which altitudes is the rate of change of the temperature negative?

    (c) At what altitude is the rate of change of the temperature largest?

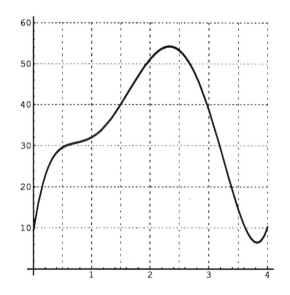

temperature (°C) vs. altitude (km)

28. The population of lemmings in Norway is growing at the rate of 3% per year. The current population of lemmings is 50,000.

    (a) If $P(t)$ is the population of lemmings t years from now, give the equation of the tangent line to the population at time 0.

    (b) Use the tangent line at time 0 to approximate the population of lemmings four months from now.

    (c) Which of the following gives the best estimate of the population of lemmings three years from now?

        i. $50,000(1.09)$
        ii. $50,000(1.03)^3$
        iii. $50,000(1.01)^3$
        iv. $50,000(1.0025)^{36}$
        v. $50,000(1.01)^9$

29. The two graphs below give the *marginal tax* on taxable income.

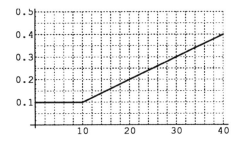
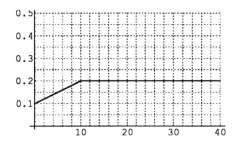

Tax A: marginal tax (%) vs. income ($1000)

Tax B: marginal tax (%) vs. income ($1000)

(a) Under which scheme would a taxpayer pay the most tax on a taxable income of $20,000? Justify your answer.

(b) Under which scheme would a taxpayer pay the most tax on a taxable income of $35,000? Justify your answer.

30. A model for the population of birds in a wildlife refuge is given by

$$P(t) = \begin{cases} 0.1t^2, & \text{for } t \leq 10 \\ \sqrt{10t}, & \text{for } t > 10 \end{cases}.$$

where $P(t)$ gives the poulation of the birds (measured in hundreds) t years from now. Use the model to answer the following questions.

(a) How many birds will be in the refuge 40 years from now?

(b) Sketch the graph of P versus time for the next 40 years.

(c) Is P a continuous function of t when $t = 10$ years? Explain.

(d) What is the average rate of change of the population of the birds in the refuge during the period from $t = 9$ years to $t = 40$ years?

(e) At what rate will the population of birds be changing in 9 years?

(f) (**Bonus**) Sketch the graph of the rate of change of the population for $[0, 20]$.

31. A driveway needs to be treated with a chemical to preserve the asphalt. A gallon of the chemical costs $23.00 and will cover 125 square feet of the driveway. The distance from one end of the driveway to the road is 100 feet. The width of the driveway was measured every 10 feet, and the following measurements were obtained.

Distance to road	0	10	20	30	40	50	60	70	80	90	100
Width of driveway	20	18	17	20	22	24	22	25	27	28	30

(a) Use the trapezoidal rule to estimate how many gallons of the chemical will be needed to cover the driveway.

(b) Use Simpson's rule to estimate how many gallons of the chemical will be needed to cover the driveway.

(c) How many gallons of the chemical would you buy? Explain your choice.

32. Decide whether $f(x) = \int_0^x \sqrt{t^2 + t}\, dt$ has an inverse. Justify your decision.

33. A trapezoidal-shaped lot has 200 feet along the front of a highway. It is 100 feet deep and the back side (the side that is parallel to the highway) is 50 feet. The assessed value of a square foot x feet from the highway is $500 - 2x$ dollars.

    (a) Find the area of the lot.

    (b) Find the assessed value of the lot.

34. A function $f(x)$ is called a probability density function for $[0, \infty)$ if the integral of $f(x)$ from 0 to ∞ converges and is equal to 1.

    (a) Show that the function $2e^{-2x}$ is a probability density function.

    (b) Compute the integral from 0 to ∞ of $x(2e^{-2x})$. (This is called the mean of the distribution.)

35. True or false? If you answer false, give a counterexample for full credit.

    (a) If $\lim_{n \to \infty} a_n = 0$, then $\sum_1^\infty a_n$ converges.

    (b) If $a_n > 0$ for all n, then $\sum_1^\infty a_n$ diverges.

36. Let $P(t)$ represent the price of a share of stock at time t. What does each of the following statements tell you about the behavior of the graph of $P(t)$? Be as precise as you can.

    (a) The price of the stock is rising faster and faster.

    (b) The price of the stock is close to bottoming out.

Appendix C

Guidelines for Projects

Projects are an important component of this course. You will be expected to complete three projects this semester. Each project will present you with a substantial problem to solve. Solving the problem will take two or three weeks. You will be working in assigned groups of three or four students. All the members of your project group will receive the same grade for the group portions of the project. Here are some suggestions:

1. **Group work**: Be sure you get off to an early start, since projects require extensive thought and development of ideas as well as clear, concise writeups. Your group should meet shortly after the project is assigned to map out a solution strategy and several more times during the actual solution period. Often when a group is writing up its report, someone finds that there is an error in the proposed solution. Therefore, you should aim to have your report completed well before it is due.

    It is important that everyone in the group participates in the work of the group. In particular, everyone needs to understand how the problem is being solved. To insure this, you should rotate the role of leader/secretary of your group. Be aware that *any* group member may be asked to report on the group's progress or final results.

2. **Consultations**: You should feel free to consult me about your projects. I will try to point you away from undue difficulties without giving away the heart of the project. Some consultations will be required, especially for the early projects.

3. **Formal writeup**: Your final report is to be typed on standard $8.5 \times 11''$ paper. Equations and graphs may be neatly hand written. Graphs are to be clearly drawn and well labeled or computer produced. Be sure that the names of all the members of the group appear on the cover page.

In writing your report, assume the reader is a student in another calculus class who has not worked on this project. Annotate any derivations that appear in the report, and explain the steps in your reasoning.

Take as much pride in your report as you would if you were writing it for an employer on whom you wish to make a favorable impression.

4. **Free reading**: If you submit your project report to me at least three days before it is due, I will read it to detect major misconceptions and return it to you for revision before the due date. This service will not lower your grade.

5. **Efficiency**: Here are some suggestions to help you work more efficiently. As soon as the group is finished with a part of the project you should write that part of the report. This will help you avoid the "all nighter" syndrome the night before the project is due. Word processing will be a big help since it makes it easy to make changes (e.g., after an early submission). Word processing also makes it possible for different people to share in the typing of the report. You should *avoid* a group setup where one person does the "thinking" and a different person is responsible for "production" of the report. If you don't know any word processing see me. I am assuming you all have some experience with word processing. If not, I will have Academic Computing Services present a short orientation for the class. [Note: in Calculus 1 we schedule an introduction to word processing for the class.]

6. **Meetings**: Meetings should have some structure and a time limit. You should each think about the project *before* the meeting. After two hours almost any meeting is much less productive. Before the end of any meeting you should decide what is to be done (and who is doing it) before the next meeting.

7. **Log**: Your group should keep a log. The log should be handed in with your final report. It should include at least the following: times the group met, members who attended that meeting, brief summary of any decisions reached (e.g., Joe will type up this part, Mary will draw a graph, we will use Wordperfect on the report, John will investigate $\alpha = 4$ for next time, etc.)

8. **Peer evaluation**: Each member will hand in an evaluation of the other members' performance with the final report. This report is confidential—I am the only person who will see it. I will not use these reports in assigning grades on the project. At the end of the semester, the totality of all peer reports is one of the inputs I will use to decide borderline grades. You may also include in the peer reports any suggestions, improvements, or comments about the project.

Appendix D

Guide to the Threads

Listed below are the "threads" that we identified, together with lists of activities and projects that relate to each.

Graphical calculus

Activities: Chalk toss, Classroom walk, Biking to school, Raising a flag, Library trip, Airplane flight with constant velocity, Graphical estimation of slope, Slope with rulers, Examining linear velocity, Given velocity graph, sketch distance graph, Function-derivative pairs, More airplane travel, Dallas to Houston, Water tank problem, Tax rates and concavity, Testing braking performance, The start-up firm, Graphical composition, The leaky balloon, Inverse function from graphs, Postage, What's continuity?, Limits and continuity from a graph, Slopes and difference quotients, Linear approximation, Using the derivative, Gotcha, Animal growth rates, Oil flow, Graphical integration, How big can an integral be?, Finding the average rate of inflation, Cellular phones, Ferris wheel, Inverse functions and derivatives, Direction fields, Using direction fields, Drawing solution curves, The hot potato

Projects: Designing a roller coaster, Tidal flows, Designing a cruise control, Designing a detector, Taxes, Mutual funds, Topographical maps

Distance and velocity

Activities: Classroom walk, Biking to school, Library trip, Airplane flight with constant velocity, Examining linear velocity, Given velocity graph, sketch distance graph, More airplane travel, Dallas to Houston, Testing braking performance, Gotcha, Time and speed, Can the car stop in time?, Cellular phones, Ferris wheel

Projects: Designing a cruise control, Designing a detector, Rescuing a satellite

Multiple representation of Functions

Activities: Water balloon, Slope with rulers, Testing braking performance, The leaky balloon, Introduction to functions, Estimating cost, Time and speed, Can the car stop in time?, Fundamental theorem of calculus, Graphical integration, How big can an integral be?, Cellular phones, Exponential differences, Log-log plots, Using scales, Spread of a rumor, Taylor series

Projects: Designing a cruise control, Designing a detector, Taxes, Water evaporation, Mutual funds, Rescuing a satellite, The fish pond, Investigating series

Modeling

Activities: Chalk toss, Classroom walk, Biking to school, Raising a flag, Library trip, Airplane flight with constant velocity, Projected image, Water balloon, More airplane travel, Water tank problem, Graphical composition, Gotcha, Magnification, Oil flow, How big can an integral be?, The shorter path, Ferris wheel, Inverse functions and derivatives, The hot potato, Spread of a rumor, Population, Save the perch

Projects: Designing a roller coaster, Tidal flows, Designing a cruise control, Designing a detector, Spread of a disease, Tax assessment, Dome support in a sports stadium, The fish pond, Drug dosage

Top-down analysis

Activities: A formula for a piecewise-linear graph, Water balloon, Given velocity graph, sketch distance graph, The product fund

Projects: Designing a cruise control, Rescuing a satellite, Dome support in a sports stadium

Approximation and estimation

Activities: Graphical estimation of slope, Examining linear velocity, Graphical composition, Inverse function from graphs, Sequences, Can we fool Newton?, Linear approximation, Estimating cost, Finite differences, Animal growth rates, Exchange rates and the quotient rule, Time and speed, Oil flow, Can the car stop in time?, Comparing integrals and series, Graphical integration, Graphical integration, Numerical integration, Verifying the parabolic rule, The shorter path, The River Sine, Why mathematicians use e^x, Exponential differences, Fitting exponential curves, Log-log plots, Using scales, Direction fields, Using direction fields, Drawing solution curves, The hot potato, Spread of a rumor, Convergence, Investigating series, Space station, Decimal of fortune, Approximating functions with polynomials, Introduction to power series, Graphs of polynomial approximations, Taylor series, Approximating logs, Using series to find indeterminate limits, Using power series to solve a differential equation, Second derivative test, Padé approximation, Using Taylor polynomials to approximate integrals, Complex power series

Projects: Designing a roller coaster, Tidal flows, Designing a cruise control, Rescuing a satellite, Spread of a disease, Tax assessment, The fish pond, Drug dosage, Investigating series, Topographical maps

NOTES

NOTES

NOTES

NOTES

NOTES

NOTES

NOTES

NOTES

NOTES

NOTES

NOTES